International Conference on Ecosystem, BioTechnology Agriculture and Environmental Science 2020

IOP Conference Series: Earth and Environmental Science
Volume 1131

Online
21-23 July 2020

Editor:

Robbi Rahim

ISBN: 978-1-7138-6988-7
ISSN: 1755-1307

Printed from e-media with permission by:

Curran Associates, Inc.
57 Morehouse Lane
Red Hook, NY 12571

Some format issues inherent in the e-media version may also appear in this print version.

Institute of Physics
Dirac House, Temple Back
Bristol BS1 6BE UK

Phone: 44 1 17 929 7481
Fax: 44 1 17 920 0979

techtracking@iop.org

Additional copies of this publication are available from:

Curran Associates, Inc.
57 Morehouse Lane
Red Hook, NY 12571 USA
Phone: 845-758-0400
Fax: 845-758-2633
Email: curran@proceedings.com
Web: www.proceedings.com

TABLE OF CONTENTS

AGRICULTURE

IOP Conf. Series: Earth and Environmental Science 1131 (2023) 011001 doi:10.1088/1755-1315/1131/1/011001

International Conference on Ecosystem, Biotechnology agriculture and Environmental Science

Preface

We are very happy to bring you the proceedings of the first International Conference on Ecosystem, Biotechnology agriculture and Environmental Science conjunction with International Conference on Agriculture and Environmental Science. This conference is intended to maintain knowledge exchange on policy support that allows for improving resource management for agricultural competitiveness towards progressive, self-reliance, and modern agriculture. The conference was held virtually, hosted by the Forum Kerjasama Pendidikan Tinggi and UPN "VETERAN" JAWA TIMUR.

Agriculture and rural sector remain to be major economic drivers at which agricultural producers and rural economies need to be supported through appropriate policies. Policy support should allow a fair agricultural product competitiveness, both for local markets and for global trade. With all challenges faced by stakeholders in rural areas, the availability of resources needs to be inclusively managed to meet the right balance between economic growth, poverty reduction, and environmental sustainability. High recognition on this condition would lead to key elements as foundation to recreate appropriate policy, institutional role, and financial support to promote agricultural resource competitiveness for sustainable and inclusive development of rural sector.

With expressing experiences and analyses from different angles, various sound policy recommendation in respect to the conference theme have been drawn for the interest of all concerned. The outbreak of coronavirus since 2020 has been vastly scattered and extremely influence almost all aspects of human life. Agricultural sector has also been affected with quite large of disadvantages. Concern on the current health situation and its impact in agricultural sector, the influence of pandemic COVID-19 has also been raised in any of the topics of this conference. Its impact on efforts to accelerate improvement of agricultural competitiveness would be valuable in addressing agricultural sustainability.

Holding a conference during the COVID-19 pandemic requires a series of hard work to bring the most advanced technology at all stages of the conference. The International Conference on Ecosystem, Biotechnology agriculture and Environmental Science event is virtually implemented with a model that all invited speakers are given time to present their material for about 30-40 minutes each. It then followed by a question and answer by the participants with a direct questioning system, and through chat feature with Q&A forums provided by the zoom application. Moreover, the presentations were organized through recorded materials followed by live Q&A along with chat feature of the virtual meeting.

We would like to extend our sincere thanks to all Science Committee members who come from Indonesia, Malaysia, Philippines, as well as authors who contribute significant thoughts and ideas through the papers written in collaboration with respective counterparts not only from Indonesia but also from different countries, such as those from Malaysia, Philippines, Thailand, and USA.

We also want to thank the publisher for publishing the proceedings. May the readers could enjoy the gain some valuable knowledge from it. We are expecting more and more experts and scholars from all over the world to join this international event in the future.

Editors

Robbi Rahim

Host Organizer

Forum Kerjasama Pendidikan Tinggi
UPN Veteran Jawa Timur

Peer Review Statement

All papers published in this volume have been reviewed through processes administered by the Editors. Reviews were conducted by expert referees to the professional and scientific standards expected of a proceedings journal published by IOP Publishing Publishing.

- **Type of peer review:** Double Anonymous
- **Conference submission management system:** Morressier
- **Number of submissions received:** 67
- **Number of submissions sent for review:** 67
- **Number of submissions accepted:** 24
- **Acceptance Rate (Submissions Accepted / Submissions Received × 100):** 35.8
- **Average number of reviews per paper:** 2
- **Total number of revlewers Involved:** 10
- **Contact person for queries:**
  Name: Robbi Rahim
  Email: robbirahim@ieee.org
  Affiliation: Sekolah Tinggi Ilmu Manajemen Sukma

IOP Conf. Series: Earth and Environmental Science 1131 (2023) 012001 doi:10.1088/1755-1315/1131/1/012001

Strengthening Farmer Organization for Smallholder Farmers

Dhina Mustikaningrum*

Universitas Sunan Bonang, Tuban, Indonesia

*dhina.mustikaningrum@gmail.com

Abstract. One of Smallholder farmers face various obstacles in developing their farming business including access to capital one of the efforts to answer the challenge is to strengthen farmer organizations. In Tuban, there is a HIPPA or Water User Farmers Association that is able to provide capital for farmers and contribute to Village Income. This study aims to analyze strategies to strengthen HIPPA Sekar Padi in Bandungrejo Village to increase capital loans for farmer and village income. The method used in this study is SWOT analysis based on the results of processing organizational documents and in-depth interviews with HIPPA administrators and farmers. The main strategy to strengthen HIPPA to contribute to village income and provide capital for farmers is reduce fuel and electricity consumption and ensure that the Tuban Regency Government accelerates the construction of flood prevention infrastructure. Through changing the electricity tariff class from business to industry, HIPPA can contribute 35-45 million rupiah to village income in dry season and more than 100 million rupiah to increase capital for famers. In the rainy season, HIPPA relies heavily on government policies to control floods. By increasing capital for farmers, it is hoped that it can help farmer access production facilities. By additional village income, the village government can build basic infrastructure and increasing social funds, so that the welfare of the smallholder farmers as part of SDG's indicator can be improved.

1. Introduction

Tuban Regency is known as the national rice barn with an area of 56,479 hectares of rice fields(1). This is because Tuban is crossed and is directly adjacent to the Bengawan Solo River which is a provider of irrigation water throughout the year. However, the people of Tuban Regency are not yet prosperous. Based on poverty statistics in Tuban Regency released by the Central Statistics Agency in 2021, as many as 17.51% of its citizens are poor. They are generally smallholder farmers.

Smallholder farmers are farmers who have a maximum of 2 hectares of cultivated land(2). Although the term smallholder has not been officially used for categorization, in Indonesia, smallholders are considered as farmers who manage land less than 0.5 hectares(3). The main focuses of their farming activities is to meet their own family's food needs, rely on labor in the family and are subsistence oriented, so that the capital used is not too large.

Smallholder farmers face various obstacles in developing their farming business. Farmers find it difficult to deal with the low price of rice harvest, while the price of production facilities including fertilizer is increasing. This is experienced by small farmers in Bandungrejo Village, Plumpang District, Tuban Regency. This condition forces local farmers to access sources of capital. In fact, the government has supported small farmers in obtaining access to capital through various schemes, although some farmers are faced with administrative requirements. An alternative for capital financing for farmers in Bandungrejo village is a special loan for members from the Sekar Padi Water-Using Farmers Association (HIPPA) in Bandungrejo Village.

HIPPA is a farmer organization that runs the business of distributing irrigation water from the Bengawan Solo River. HIPPA will get profit sharing from farmers for this irrigation service. Profit sharing from farmers in the form of grain is then sold by HIPPA as the main source of their income. As an organization that is also engaged in the social sector, HIPPA is able to contribute to Village Income (PADes), social funds and loan funds to its members through the capital loan sub-unit. So that HIPPA can be an option in providing capital for small farmers. This study aims to analyze strategies to strengthen HIPPA Sekar Padi in Bandungrejo Village as an organization whose existence is very important for smallholder farmers.

2. Material and Methods

2.1. Material

This study was conducted in Bandungrejo Village, Plumpang District, Tuban Regency during July-August 2021. This location was chosen due to the availability of data during the planting season in the last three years, from 2018 to 2020. The materials needed for data collection related to this study are a questionnaire form for farmer in-depth interview and the HIPPA Sekar Padi Bandungrejo accountability report from 2018 to 2020.

2.2. Methods

The data used in this study are secondary data and primary data. Primary data was obtained through an in-depth interview process with the management and members of HIPPA Sekar Padi in Bandungrejo Village. While the secondary data used is the HIPPA Sekar Padi Accountability Report in Bandungrejo Village. All data were analyzed using the SWOT (Strength, Weakness, Opportunities, Threat) method to obtain a strategy for strengthening the HIPPA organization.

3. Results and Discussion

3.1. Profile of HIPPA Sekar Padi Bandungrejo

HIPPA Sekar Padi Bandungrejo was formed in 1995 as part of the follow-up to the construction of irrigation facilities assisted by the JICA (Japan International Cooperation Agency). Currently, the area of rice fields managed by HIPPA is 235 hectares in Bandungrejo Village. There are 428 HIPPA members. In one year, HIPPA serves irrigation twice, once in the rainy season and once in the dry season (Figure 1a).

The income obtained by HIPPA Sekar Padi in Bandungrejo Village comes from the sale of grain from profit sharing with farmers as members of the organization. Profit sharing from farmers is set at 14% of the farmers' net harvest. HIPPA net income is obtained after deducting operational costs, management fees, and honorary services including village development funds and social funds. To increase revenue, HIPPA seeks to reduce operating expenses, which are mostly sourced from fuel and electricity costs (Figure 1b).

The net income obtained by HIPPA Sekar Padi Bandungrejo from 2018 to 2020 there is always a gap between the dry season and the rainy season. Technically, the biggest operational costs in the dry season are the use of fuel and electricity. Meanwhile, in the rainy season, the biggest operational cost that must be allocated by HIPPA is flood management. On the other hand, rice yields generally decrease in the rainy season due to pest attacks and flooding. This decrease in rice yields has an impact on farmers' income in the rainy season which is reduced by up to 25% compared to the dry season(4). Another study stated that losses due to flooding were once estimated at 6.5-7 million per hectare(5).

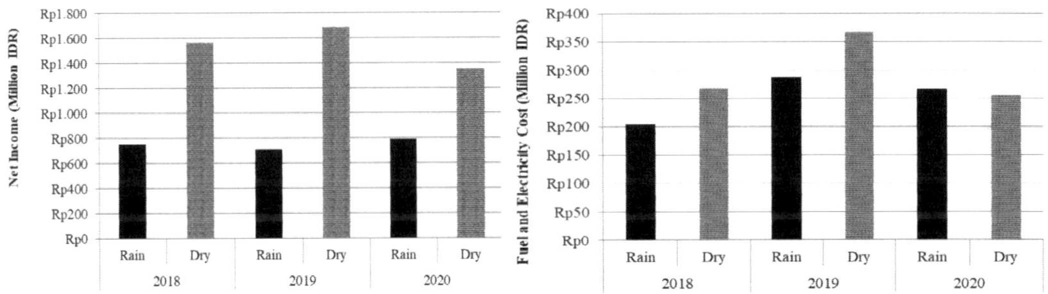

(a) (b)

Figure 1. Net Income of HIPPA Sekar Padi Bandungrejo (a) and Fuel and Electricity Consumption (b)

Based on the business model and service offering, HIPPA can be categorized as a supply organization. The characteristic of this model are having orientation in resources, Community-based, and focused on acquiring inputs for efficient subsistence and small-scale production(6). Therefore, HIPPA always contributes to the village's original income and supports capital for member farmers. This is very helpful for farmers because farmers generally do not access funding from formal institutions. Their sources of capital come from informal institutions such as Bank Harian, moneylenders, rice traders, traders in production facilities, and rice mills(7). Farmer's credit on capital is proven to affect rice production because it affects the level of technology adoption which has an impact on increasing production(8).

The amount of capital distributed by HIPPA to farmers in 2018 reached 198 million IDR, then decreased in 2019 and 2020 to 186 million IDR (Figure 2a). This decrease occurred because some farmers have not been able to repay their capital loans. Even though this loan is relatively light because there is no interest and the farmer is only charged an administrative fee. Farmers borrow capital because they still lack capital, the cost of living is high, and they depend on their agricultural products to meet their daily needs(9).

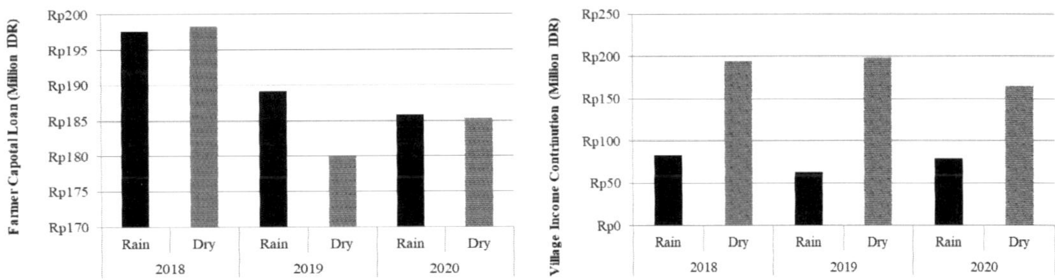

(a) (b)

Figure 2. HIPPA contribution to capital loans (a) and village income (b)

The amount of capital distributed by HIPPA to farmers is determined through member meetings and adjusted to HIPPA's income and cash in the current year. So if HIPPA is to increase capital for farmers, HIPPA's income and cash must be high. This is a challenge for HIPPA because HIPPA operations cannot be suppressed, especially in the consumption of fuel oil and electricity. This is different from determining the value of HIPPA's contribution to village income which is determined between 10-12 percent of net income (Figure 2b).

3.2. Strategy to strengthen HIPPA Sekar Padi Bandungrejo

Based on in-depth interview result, the main strength as an internal factor to develop strategy for strengthening HIPPA is big contribution to village income and provides capital for the members. The internal weakness of the organization is the big consumption of fuel and electricity as part of organizations operational. Externally, a great opportunity from HIPPA to increase income is to develop service areas outside the village of Bandungrejo. The main threat is the regular occurrence of floods which impact on the decreasing crop yields and organizational income.

Therefore, if HIPPA wants to increase its contribution to village income and farmer's capital, then the efforts that must be developed by HIPPA are to reduce fuel and electricity consumption and ensure that the Tuban Regency Government accelerates the construction of flood prevention infrastructure. Currently, the rate charged to HIPPA is business class. Based on state electricity company (PLN), business tariff is 23% higher than industry class. In-depth interview result concluded that HIPPA has to negotiate with state electricity company to change electricity tariff from business class to industry class (Table 1).

Related to fuel, HIPPA always makes purchases at subsidized prices. This is considered a violation of the rules because the type of business HIPPA is commercial, whereas HIPPA is also a social organization. HIPPA management is trying to collaborate with Pertamina in order to have a special gas station for farmers as well as a special gas station for fishermen that has been developed by Pertamina. Special gas station for fishermen is called the Fisherman's Fuel Filling Station because it serves fishermen and fishery businesses (10). HIPPA must collaborate with the government in efforts to change electricity rates and build gas stations specifically for farmers. The government must also cooperate with the community in managing the productive irrigation potential(11).

The results of the interview with the management concluded that through the implementation of the strategy for changing the electricity tariff class, it is estimated that HIPPA operational costs will decrease by 16%. HIPPA can add 35-45 million rupiah to the contribution to village income if the operational cost reduced more than 15%, so that the total contribution becomes more than 200 million in the dry season. HIPPA is also able to add 75-100 million rupiah to be offered to members if the electricity and fuel costs can be reduced and increase capital for famers up to 297 million rupiah. In the rainy season, HIPPA relies heavily on government policies to control floods.

By increasing capital for farmers, it is hoped that it can help farmer access production facilities, such as fertilizers, and efforts to control pests and diseases of rice plants. The addition of village income is expected to assist the village government in building basic infrastructure and social funds, so that the welfare of the community, especially smallholder farmers, can be increased.

4. Conclusion

The main strategy to strengthen HIPPA to contribute to village income and provide capital for farmers are reduce fuel and electricity consumption and ensure that the Tuban Regency Government accelerates the construction of flood prevention infrastructure. Through changing the electricity tariff class from business to industry, HIPPA can contribute 35-45 million rupiah to village income in dry season and more than 100 million rupiah to increase capital for famers. In the rainy season, HIPPA relies heavily on government policies to control floods. By increasing capital for farmers, it is hoped that it can help farmer access production facilities. By additional village income, the village government can build basic infrastructure and increasing social funds, so that the welfare of the farmers as part of SDG's indicator can be improved.

Table 1. SWOT Matrix

External Factor / Internal Factor	**Opportunities/O (External)**	**Threats/T (External)**
	• Irrigation water available all year round • Can develop business beyond administrative boundaries • Cooperation with farmers group (Gapoktan) • Cooperation with government	• Flood events on a regular basis • Fluctuating rice harvest prices
Strengths/S (Internal) • Big contribution to village income • Contribution to farmer's capital • Have a lot of organization asset • Big number of organization member	**S-O** • Legal entity to save the asset and is recognized by the state • Increasing the value of farmers' capital loans • Increasing village income by developing business to other village	**S-T** • Ensuring the involvement of HIPPA in the flood management process facilitated by the government
Weaknesses/W (Internal) • Human resource of the HIPPA management, mostly have the last education in junior high school • Conventional management model • Big fuel and electricity consumption	**W-O** • Management training by government for HIPPA administrator (including reporting skill, negotiation skill, leadership and communication) • Assistance in developing village agreement (bussiness development) • Assistance in changing the electricity tariff category from business to industry (negotiate with state electricity company) • licensing in building special gas stations for farmers	**W-T** • Accelerates the construction of flood prevention infrastructure

Source: Processed Data (2021)

5. References

1. Badan Pusat Statistik. Kabupaten tuban dalam angka 2020. 2021;
2. Syahyuti. Eksistensi dan Esensi Peran Pertanian Skala Kecil dalam Pemenuhan Pangan Nasional: Studi Kasus Negara-Negara ASEAN. :485–500.
3. Susilowati SH, Maulana M. Luas Lahan Usahatani dan Kesejahteraan Petani : Eksistensi Petani Gurem dan Urgensi Kebijakan Reforma Agraria Farm Business Land Size a nd Farmers ' Welfare : Smallholders ' Existence and Agrarian Reform Urgency Lahan pertanian dewasa ini menghadapi tanta. Anal Kebijak Pertan. 2012;10(1):28.

4. Roidah IS. Analisis Pendapatan Usahatani Padi Musim Hujan Dan Musim Kemarau (Studi Kasus di Desa Sepatan Kecamatan Gondang Kabupaten Tulungagung). Agribis. 2015;11(13):45–55.
5. Makarim AK, Ikhwani I. Inovasi Dan Strategi Untuk Mengurangi Pengaruh Banjir Pada Usahatani Padi. J Ilmu Tanah dan Lingkung. 2011;13(1):35.
6. Hansen AR. Improving Institutional Capacity : Strengthening Farmer Organizations In the OIC Member Countries. 2015.
7. Supriatna A. Aksesibilitas Petani Kecil Pada Sumber Kredit Pertanian di Tingkat Desa: Studi Kasus Petani Padi di Nusa Tenggara Barat. SOCA J Sos Ekon Pertan. 2008;8(2):0–15.
8. Wati DR, Nuryartono N, Anggraeni L. Akses dan Dampak kredit Mikro Terhadap Produksi Padi Organik di Kabupaten Bogor. J Ekon dan Kebijak Pembang. 2014;3(2):75–94.
9. Supanggih D, Widodo S. Aksesibitas petani terhadap lembaga keuangan (Studi kasus pada petani di desa Sidodadi kecamatan Sukosewu kabupaten Bojonegoro). J Agriekonomika. 2013;2(2):163–73.
10. Pertamina. Kebijakan Pembangunan SPBN/SPDN Dalam Memenuhi Kebutuhan BBM Bagi Nelayan. 2016.
11. Supriadi H, Rivai RS. Pengembangan Investasi irigasi Kecil Untuk Pemingkatan Produksi Padi Mendukung Swasembada Beras. Anal Kebijak Pertan. 2018;16(1):43–57.

IOP Conf. Series: Earth and Environmental Science 1131 (2023) 012002 doi:10.1088/1755-1315/1131/1/012002

Agronomic performance of F7 soybean lines resistant to *Soybean mosaic virus* in the dryland area

Wuye R. Andayanie[1*], Praptiningsih G. Adinurani[1], Martin Lukito[1], Akas Y.Pulihasih[2]

[1]University of Merdeka Madiun, Madiun, Indonesia.
[2]Nahdlatul Ulama Surabaya University, Surabaya, Indonesia.

*wuye.andayanie@gmail.com

Abstract. In Indonesia, soybean cultivation is generally planted in paddy fields after the rice harvest. However, expanding soybean cultivation is needed to make optimal dry land in the future. This study aims to obtain F7 soybean lines resistant to *Soybean mosaic virus* (SMV) adaptive in the dry land. In the dry land planted eleven soybean breeding lines resistant to SMV and two varieties as control. Ten days after planting (DAP), soybean plants were inoculated with SMV. The experiment was conducted in the randomized complete block design with four replicates within each area. These results showed that the GK/PI200.485-7-17 line was the most superior soybean line and the potency to develop in the dryland area because it had the highest number of branches per plant, number of filled pods per plant, and 100-seed weight. The tested line showed the highest seed yields of 2.40 t/ha, and 2.72 t/ha, in Girisubo and Sidowayah Districts, respectively. This line did not indicate any serological relationship to SMV in both locations.

1. Introduction

Soybeans contain an essential source of protein for food and feed. Soybean cultivation is generally planted after rice planting in the dry season. Indonesia's total upland area with a dry climate had reached 99.65 million ha that potential land for agriculture. In comparison, the remaining 44.82 million ha is not the potential for agriculture lies primarily in Forest. Increasing productivity and expansion of the planting area increased soybean production. Maize and soybean plants usually become a priority crop for planters during the rainy season in the dryland areas. Therefore, soybean productivity in high rainfall was higher than in low [1, 2].

The soybean plantation can adapt to Grumusols, Inceptisol, Ultisols, Alfisol soil types. However, productivity decreased due to the use of the intercropping system [3]. Another study showed that soybean planting could adapt well on dryland with a dry climate and implement under the teaks to meet the domestic demand for soybeans. In the forest area, soybean cultivation has caused problems related to productivity, i.e., soil fertility, biophysical conditions, and competition with non-food crops [4]. *Aphis glycines* can be an important pest of young soybean plants if there is insufficient rainfall in dryland crops. These aphids also transmit the *Soybean mosaic virus* (SMV). The SMV causes mosaic, leaves crinkle, stunting of the plant [5].

Moreover, SMV could decrease seed quality (mottle and deformation of seeds). The symptoms varied depending on host plant genotype, the strain of virus, environmental conditions, and plant age at the time of infection. Some soybean lines are more tolerant to the SMV. Botanical antiviral and insecticide could reduce mosaic disease and aphids, respectively. Cashew nutshell extract contains anacardic acid, which has a dominant role as an active antifeedant compound [6].
One of the crucial stages in the assembly of new varieties dryland tolerant and resistant to SMV is a test of soybean lines at several locations. The assembly of lines can increase soybean production on dry land and resistance to SMV. The growth stability and product lines in several areas carried out one crucial step to determine prospective lines at several locations. Some obstacles to acid dryland management are acidic soil, phosphorus content, calcium, microorganism activity, soil fertility, high levels of aluminum (Al), zinc (Zn), and iron. Susceptible soybean lines have root growth inhibition under dryland conditions [7]. Soil is an essential source of plant nutrients. Therefore, soybean plants can adapt to dry land if the plant can absorb nutrients and reduce the transpiration rate.
Managing dryland resources include promoting integrated pest and soil fertility management. Rhizobia inoculation has a significant effect on land with low nutrient content and water availability. The use of seed treatment with Rhizobium played a crucial role in enhancing nitrogen fixation of legume crops and productivity. Moreover, planting-virus free seed and resistant varieties can also reduce *Soybean mosaic disease* [8, 9]. Therefore, this research aimed to find out the resistance to SMV in the dry land of F7 soybean lines.

2. Material and Methods

This research was carried out in Plant Protection Laboratory, Department of Agrotechnology, University of Merdeka Madiun - Indonesia. The field experiments were initiated from the end of January to early April 2021 during the rainy season. Soybean planting was grown in Sidowayah District, Ngawi Regency (East Java Province- Indonesia) at a temperature range of 28.6 to 29.8 °C with a relative humidity range of 75 to 80%, while in Girisubo District, Gunung Kidul Regency (Central Java Province-Indonesia) at a temperature range of 23.8 to 26.3 °C with a relative humidity range of 80 to 85%. Sidowayah and Girisubo District have average monthly rainfall of (102 mm, 70 mm) from January to April with average monthly rainy days of (11 d, 15 d) during the growing season, respectively (Table 1).

Table 1. Rainfall and rainy days during the periode January to April, 2021

Month/Period	Sidowayah District		Girisubo District	
	Average monthly rainfall (mm)	Average monthly rainy days (days)	Average monthly rainfall (mm)	Average monthly rainy days (days)
January	143	9	108	12
February	97	16	70	19
March	120	11	77	24
April	48	8	25	5

Note: Source Meteorological, Climatological, and Geophysical Agency (MCGA) in Ngale and Yogyakarta station.

All of the soils in both locations have Alfisol soil type. One of the experiments belonged to the highly acidic soil conditions range (i.e. pH 4.7 to 5.3), and another experiment was a slightly basic soil conditions range (i.e. pH 6.3 to 6.8). Table 2 showed the soil chemical properties in field experiments.

Table 2. The soil chemical properties of dryland in field experiments

Chemical properties	Value	
	Girisubo District	Sidowayah District
Soil types	Alfisol	Alfisol
pH H_2O	4.7 to 5.3	6.3 to 6.8
C-organic (%)	0.42 to 1.55	4.87
N (%)	0.09 to 0.015	0.02 to 0.04
Bray-P1 (ppm)	-	1.55 to 11.74
Olsen-P (ppm)	18.25 to 44.01	6.41
K (me/100 g)	0.19 to 0.55	0.72
Ca (me/100 g)	0.81 to 5.83	6.59
Mg (me/100 g)	0.55	0.62 to 1.96
Fe (ppm)	12.45 to 51.98	1.99
SO_4 (ppm)	12.71 to 31.87	17.10
Mn (ppm)	37.22 to 248	0.00 to 0.11

Note: The result of soil analysis of Physics and Chemistry Laboratory, Faculty of Agriculture, Gadjah Mada University (2021).

2.1. Material

Soil preparation. Soybean planting is done with 2.4 × 4.5 m and 38 x 12 cm plant spacing. The number of each clump is uniform to two plants. Botanical herbicide is applied on the soil surface before using the tillage system. The viscous extract of cashew nutshell was diluted with water to obtain a concentration of 10% when sprayed, mixing with surfactants at 100 ml/1000 ml of water [10]. Soil tillage systems were applied by ploughing the soil and level with a rake. Fertilization was carried out with essential fertilizers of Urea (75 kg/ha), SP-36 (100 kg/ha), and KCl (100 kg/ha). *Pseudomonas fluorescens* as phosphate solubilizing activity (5.25 L/ha) and farmyard manure (100 kg/ha) were applied to soil at five days after planting (DAP).

Soybean seed treatment preparation. The experiment used a soybean resistant genotype consist of 11 soybean lines resistant to the *Soybean mosaic virus* and two check varieties (Dering 1, Anjasmoro). Dering 1 and Anjasmoro are varieties with dryland and shading tolerance, respectively. Soybean seed variety and F7 Soybean Lines seed were derived from Indonesian Legumes and Tuber Crops Institute (ILTCRI) and Faculty of Agriculture, University of Merdeka Madiun - Indonesia, respectively. F7 Soybean lines were obtained by the modified bulk method. Soybean seed (0.5 kg) was mixed cold rice gruel (250 mL) and rhizobial isolate (2.5 g). The mixing procedure should be done with care. The coated seeds were air-dried for 30 min before sowing.

Soybean mosaic virus. The *Soybean mosaic virus* (SMV) isolate was obtained from soybean plant propagation that showed mosaic symptom and serological relationship to SMV. Ten DAP, soybean plants were inoculated with SMV by the compressed air technique (1.8 to 2.0 kg /cm^2). Inoculum (1 g of diseased leaves/50 mL of buffer) and carborundum 600 mesh 0.5% (0.5 g/100 mL) solution was sprayed about 10 to15 cm from the plant surface for 0.5 s/plant.

2.2. Methods

Inoculation of *Soybean mosaic virus*. The *Soybean mosaic virus* (SMV) isolate was obtained from soybean plant propagation that showed mosaic symptom and serological relationship to SMV. Ten days after planting (DAP), soybean plants were inoculated with SMV by the compressed air technique (1.8 to 2.0 kg/cm^2). Inoculum (1 g of diseased leaves /50 mL of buffer) and carborundum 600 mesh 0.5% (0.5 g/100 ml) solution was sprayed about 10 to15 cm from the plant surface for 0.5 s/plant.

The disease incidence and virus detection. The disease incidence (DI) is measured from percentage of the total plants infected with virus to the total of plant assessed in each replicate plot on each soybean line and variety. The percentage of DI was calculated using the following formula:

$$DI = \frac{a}{N} \times 100\%$$

Where

DI : the percentage of disease incidence

a : number of plants with SMV symptom in each plot

N : number of plants in each plot

Virus detection was carried out on plant leaves four weeks after inoculation. The samples were confirmed by Indirect ELISA with the modified Koenig method and using the polyclonal antibody SMV. Virus accumulation was quantified at a wavelength of 405 nm by using ELISA Reader [11].

Statistic. A total of 11 soybean genotypes resistant to *Soybean mosaic virus* (W/Mlg 3288-7-9, W/L.Temanggung-7-1, W/L.Temanggung-7-14, GK/L.Temanggung-7-9, W/L.Temanggung-7-14, GK/L.Temanggung-7-18, GK/PI 200.485-7-2, GK/PI200.485-7-8, GK/M8Grb 44-7-1, GK/PI200.485-7-17, GK/PI 200.485-7-12) and two commercial varieties (Dering 1 and Anjasmoro) as control were selected in experimental research. The experiments were evaluated using a completely randomized blocks design (RCBD) and replicated three times within each area. A difference between lines on each variable used the F test at the 5% significant level and calculated the Least Significant Difference (LSD).

3. Results and Discussion

The difference in place affected the harvesting age of soybean and plant height. In addition, climatic conditions affected harvesting time. The harvesting time of soybean genotypes at the Girisubo District location was eight days longer than the Sidowayah District location. The harvesting age of the soybean genotype is closely related to the flowering period. So it can be seen how long it takes for soybean lines and varieties to fill seeds and determine harvesting time. Other studies have shown that the potassium (K) content in soil plays a role as a catalyst for flour, sugar, and fat formation [12]. Therefore, it can improve yield quality in flower formation, filled pods, and soybean plants' physical pods (Table 1). In addition, the mean daily rainy days has more effect on the ripening period of seeds than the mean monthly rainfall. The duration of sunlight in Girisubo District is shorter than in Sidowayah District. Thus, the number of rainy days could delay the ripening time of the seeds. However, soybean plants are more suitable for growing in Sidowayah District with high soil organic carbon (C-organic) (Table 2). Sunlight and temperature that did not follow optimal water availability and nutrients could increase virus symptoms test plants. In addition, the virus requires an active metabolic result from the host for its propagation [13].

The differences between location and genotypes influenced the agronomic performance of F7 soybean lines, i.e. the number of nodes per plant, the number of filled pods per plant, the 100-seed weight, and seed yield (Table 3).

Table 3. The Mean square values of agronomic performance in two locations of dryland areas

SV	Df	NB	PH	NN	NP	SW	SY
Environmental (E)	1	27.62**	26,184,70**	418. 08**	18,312,56**	316. 21**	25.12**
Genotypes (G)	13	1.96**	93.81	14.85**	529.14**	2,31*	0.38**
G x E	13	0.81	99.13	17.43	235.60	4.27**	0.35**
SE	69	0.77	53.81	2.72	94.72	0.93	0.05
CV (%)		26.13	11.76	10.50	15.68	7.81	15.02

Note: SV: source of variation; SE: standard error; Df: degree of freedom NB : the number of branches per plant; PH: plant height; NN: the number of nodes per plant ; NP: the number of filled pods per plant; SW: the 100-seed weight; SY: seed yield (t/ ha); CV: coefficient of variation.

Genotypes and environmental interaction strongly influenced crop yield. Several genotypes of soybean produced a higher number of filled pods per plant and seed yield in dryland areas with a slightly basic soil conditions range (i.e. pH 6.3 to 6.8) than a highly acidic soil pH range (i.e. 4.7 to 5.3) (Table 2 and Table 3). Agronomic performance and yield stability of F7 soybean lines at different locations indicate that these genotypes can adapt to regions with other fertility and climatic conditions. Genotypes with high yield in certain areas showed that they were only suitable in areas with special climatic or fertility conditions. Each genotype has different characteristics for interacting with the environment. Genotypes by environmental interaction (G x E) influenced the number of branches per plant separately. In this study, the genotype with the highest number of branches in one location showed consistency in other areas. However, environmental factors such as climate conditions affected the number of branches. Performance of plant height and number of branches per plant in Girisubo District was higher than performance in Sidowayah District location. In each genotype, the number of branches per plant ranges from two to four. GK/PI 200.485-7-12 line produced the most branches and showed no significant difference by GK/PI200.485-7-17 line, while W/L.Temanggung-7-9 line had the fewest branches. GK/PI 200.485-7-12, GK/PI200.485-7-17, and GK/PI200.485-7-17 line were 4.00, 3.98, and 2.56, respectively. The lowest number of branches of the W/PI 200.485-7-4 line is equivalent to the comparative Anjasmoro variety branches (2.75). The GK/PI200.485-7-12, GK/PI200.485-7-17, and GK/L.Temanggung-7-18 lines produced more the number of branches than Dering 1 variety, while W/L. The Temanggung-7-14 line is equivalent to the number of branches of Dering 1 variety (3.41) (Table 4).

Table 4. Number of branches, nodes, and pods per plant of soybean lines in two locations of dryland areas

Genotypes	Number of branches per plant	Number of nodes per plant		Number of pods per plant	
		Girisubo District	Sidowayah District	Girisubo District	Sidowayah District
W/Mlg 3288-7-9	2.87 c	13.37 c	9.22 d	36.86 k	59.36 f-i
W/L. Temanggung-7-1	2.74 cd	15.40 bc	11.69 cd	38.45 k	75.12 c-e
W/L. Temanggung-7-14	3.41 bc	17.84 b	15.81 bc	50.13 i-k	82.01 b-d
GK/L. Temanggung-7-9	2.89 c	12.81 c	13.37 c	40.11 k	65.18 e-h
W/PI 200.485-7-4	2.66 d	18.62 b	17.93 b	53.75 h-j	92.79 ab
GK/L.Temanggung-7-18	3,70 b	20.16 ab	15.53 bc	43.86 j-k	83.91 bc
GK/PI200.485-7-2	2.86 c	18.94 b	20.43 ab	53.49 h-j	82.15 b-d
GK/PI200.485-7-8	2.90 c	15.08 bc	12.85 c	48.50 i-k	70.42 c-f
GK/M8Grb 44-7-1	2.72 cd	13.59 c	12.26 c	38.05 k	74.65 c-e
GK/PI200.485-7-17	4.00 a	21.87 a	21.65 a	55.27 g-j	98.61 a
GK/PI 200.485-7-12	3.98 a	19.71 ab	11.12 cd	37.42 k	73.81 c-e
Dering 1	3.41 bc	12.86 c	12.90 c	46.81 i-k	70.59 c-f
Anjasmoro	2.66 d	20.12 ab	16.04 bc	52.92 h-j	67.96 d-g
LSD 0.05	0.81	2.21		13.49	

Note: values sharing by same letters are not significantly different among the treatments at the least significant difference (LSD) test with $\alpha = 0.05$.

Increasing the number of branches and pods in soybeans requires Phosphorus (P) nutrients from the soil. Moreover, Potassium (K) nutrients could improve the filling of pods and soybean yield, including the number of branches, fertile nodes, and pithy pods. Similarly, in a study conducted by Khanam et al. [14], Gebrehana, and Dagnaw (2020) [15], both macro-nutrients (P and K) influenced nodule formation and soybean yield. However, excessive P and K nutrient application will harm on plant growth.Therefore, the application of 175 kg/ha triple superphosphate (TSP) and 120 kg/ha Muriate of potash (MoP) was the most appropriate levels of P and K to maximize soybean yield.

The number of reproductive nodes per plant of lines and Anjasmoro variety in the Girisubo District was more significant than in the Sidowayah District, except for the GK/L. Temanggung-7-9 l, GK/PI200.485-7-2 lines and the Dering 1 variety. However, the number of reproductive nodes per plant of GK/L.Temanggung-7-9 line and Dering 1 variety were no significant difference ($p < 0.05$) in Girisubo and Sidowayah District. The difference in the number of reproductive nodes per plant for each line between locations ranges from 3 to 6 nodes. W/PI 200.485-7-4, GK/PI200.485-7-17 lines, and Dering 1 variety produced the number of reproductive nodes per plant stability in the Girisubo and Sidowayah District. Genetic and environmental factors on several lines and varieties influenced the number of nodes per plant. Genotypes by environmental interaction affected at 19 loci with 51 quantitative trait locus (QTLs). This locus has affected soybean plant height. A total of 13 QTLs contributed to improving soybean stem node numbers under varied environmental conditions [16]. As a result, The number of filled pods at the Girisubo District location is less than that produced at the Sidowayah District. Three soybean lines, namely W/PI200.485-7-4, GK/PI200.485-7-2, and GK/ PI200.485-7-12 in the Girisubo District location, produced more filled pods the control varieties. Meanwhile, all soybean lines in the Sidowayah District location had more filled pods than the comparison varieties, except W/Mlg 3288-7-9 and GK/ L.Temanggung-7-9 lines. GK/PI200.485-7-17 lines in Sidowayah District had the most pods per plant (Table 4). Soil nutrient availability, water, and sunlight influenced the formation of filled pods.

During the pod growth and filling phase, drought stress (average rainfall of 77 mm per month) in Girisubo District caused fewer pods per plant, and filling the pods is not optimal than in Sidowayah District. The lack of water in the growth and pod filling phase stimulates the decline of photosynthetic, flowering intensity, and pod development [17, 18]. Three soybean lines, namely W/L.Temanggung-7-14, W/PI 200.485-7-4, GK/PI200.485-7-17, and GK/PI 200.485-7-12 in the Girisubo District, had 100 seed weights larger than the dering-1. On the other hand, all soybean lines in Sidowayah District had 100 seed weights larger than comparison varieties (Dering-1 and Anjasmoro). However, W/Mlg 3288-7-9 and W/PI 200.485-7-4 had 100 seed weights smaller than Dering variety (Table 4). Therefore, we could suppose that genetic and environmental factors influenced the instability of seed weight on each line. Tables 1 and 3 show that the Girisubo District condition has lower rainfall and soil chemical properties than Sidowayah District. Therefore, these influenced soil fertility, plant growth, and seeds cannot develop optimally.

The 100 seed weight of soybean lines and varieties in the Girisubo district was smaller than in the Sidowayah District. On the other hand, the seed weight of the Anjasmoro variety was from 9 to10 g per 100 seeds to 10 to12 g per 100 seeds, and the Dering 1 variety was from 9 to 11 g per 100 seeds to 11 to 12 in two locations. Similarly, in a study conducted by Diptaningsari & Rivaie [19], the Anjasmoro variety was more adaptable for planting on optimal land and less ideal for growing on suboptimal land than Dering variety. The seed weight of Dering 1 varieties ranged from 10 to12 g per 100 seeds that is still equivalent to the Dering 1 variety in general. There was no change in seed weight in Dering 1 variety because Dering 1 was one of the varieties that tolerated dryland areas.

The seeds yield (t/ha) for each line in the Girisubo District location was higher than the Sidowayah District. Difference between locations results obtained varieties Anjasmoro as much as 0.61 t ha^{-1} and W/L.Temanggung-7-14 line produced the most as much as 1.7 t/ha. W/Mlg 3288-Two soybean lines, namely 7-9 and GK/PI200.485-7-17, had the highest seed yield at both locations. GK / PI200.485-7-17 is the most superior soybean line. Therefore, it has many branches, filled pods, 100- seed weight, and seed yield. There was the most consistent among all tested lines and control varieties. The interaction between genetic and environmental factors influenced plant performance and seed yield. Soybean line can produce a high yield performance in one place but low in another area (Table 5).

Table 5. The 100-seed weight and the seed yield of soybean lines in two locations of dryland areas

Genotypes	100-seed weight (g)		seed yields (t/ha)	
	Girisubo District	Sidowayah District	Girisubo District	Sidowayah District
W/Mlg 3288-7-9	9.42 h	10.35 gh	0.98 j	1.25 h-j
W/L. Temanggung-7-1	9.69 h	13.29 b-d	1.04 ij	1.96 cd
W/L.Temanggung-7-14	11.32 fg	14.72 a	2.16 bc	2.35 ab
GK/L. Temanggung-7-9	9.76 h	12.08 e-f	1.33 g-i	2.17 bc
W/PI 200.485-7-4	10.25 gh	11.59 f	1.40 f-i	1.85 c-e
GK/L.Temanggung-7-18	9.34 h	13.61 a-c	1.52 e-h	2.15 bc
GK/PI200.485-7-2	9.88 h	11.60 f	1.21 g-i	1.90 cd
GK/PI200.485-7-8	9.75 h	13.15 b-e	1.66 d-g	2.38 ab
GK/M8Grb 44-7-1	10.18 gh	12.31d-f	1.20 h-j	2.19 bc
GK/PI200.485-7-17	14.07 ab	14.96 a	2.45 ab	2.72 a
GK/PI 200.485-7-12	10.27 gh	13.10 b-e	1.45 f-h	2.46 ab
Dering 1	10.21 gh	11.62 f	1.83 c-e	2.50 ab
Anjasmoro	9.38 h	11.28 fg	1.65 d-g	1.89 cd
LSD 0.05	1.33		0.41	

Note: values sharing by same letters are not significantly different among the treatments at the least significant difference (LSD) test with $\alpha = 0.05$.

More considerable 100-seed weight is generally associated with lower seed yields, while many branches and seeds per pod produce a higher seed weight. Therefore, two lines, namely W/Mlg 3288-7-9 and GK/PI200.485-7-17, were potential because they gave stable seed yields at both locations and were higher than the comparison varieties.

Soybean lines showed the incidence of mosaic disease at the observation site, except the strain (GK/PI200.485-7-17) in Sidowayah District did not show disease incidence until the observation. The Anjasmoro variety showed the highest virus infection (42%) (Table 6).

Table 6. The disease incidence and absorbance ELISA value of soybean lines in two locations of dryland areas

Genotypes	Disease Incidence (%)		Absorbance ELISA Value[a]/Reaction	
	Girisubo District	Sidowayah District	Girisubo District	Sidowayah District
W/Mlg 3288-7-9	14/50 (0.28)	11/50 (0.22)	0.286 ± 0.962 a/+	0.117 ± 0.722 a/+
W/L. Temanggung-7-1	11/50 (0.22)	8/50 (0.16)	0.319 ± 0.731 a/+	0.082 ± 0.021 b/–
W/L.Temanggung-7-14	7/50 (0.14)	5/50 (0.10)	0.078 ± 0.019 b/–	0.070 ± 0.011 b/–
GK/L. Temanggung-7-9	11/50 (0.22)	6/50 (0.12)	0.214 ± 1.430 a/+	0.079 ± 0.024 b/–
W/PI 200.485-7-4	7/50 (0.14)	4/50 (0.08)	0.076 ± 0.015 b/–	0.070 ± 0.009 b/–
GK/L.Temanggung-7-18	8/50 (0.16)	5/50 (0.10)	0.080 ± 0.025 b/–	0.072 ± 0.012 b/–
GK/PI200.485-7-2	15/50 (0.30)	4/50 (0.08)	0.512 ± 0.470 a/+	0.095 ± 0.018 b/–
GK/PI200.485-7-8	6/50 (0.12)	4/50 (0.08)	0.075 ± 0.012 b/–	0.070 ± 0.009 b/–
GK/M8Grb 44-7-1	8/50 (0.16)	3/50 (0.06)	0.083 ± 0.015 b/–	0.072 ± 0.011 b/–
GK/PI200.485-7-17	2/50 (0.08)	0/50 (0.00)	0.071 ± 0.012 b/–	0.068 ± 0.006 b/–
GK/PI 200.485-7-12	9/50 (0.18)	9/50(0.18)	0.080 ± 0.022 b/–	0.079 ± 0.022 b/–
Dering 1	6/50 (0.12)	4/50 (0.08)	0.084 ±0.019 b/–	0.077 ± 0.016 b/–
Anjasmoro	16/50 (0.32)	8/50 (0.16)	0.325 ± 0.722 a/+	0.081 ± 0.017 b/–

[a] : mean obtained of pooled 10 sample plants with five replications wells; A positive reaction is determined if the Absorbance ELISA Value (AEV) of the test sample is twice the AEV of the negative control; Values sharing by same letters are not significantly different among the treatments at the least significant difference (LSD) test with $\alpha = 0.05$.

However, healthy plant had a higher population than the virus-infected plant. The soybean line has different responses at the same and other environmental conditions. The early vegetative phase had a relatively low to moderate disease incidence rate, namely a mild mosaic on few leaves. The incidence of mosaic disease increased according to age, where the disease incidence was the highest achieved at 36 DAP. The soybean genotype gave various responses due to the line, environment, and interaction between infected genotype × environment. Soybean crops with high rainfall caused a lower disease incidence than crops with low rainfall. Based on precipitation, soybean lines and varieties tended a higher disease incidence in Girisubo District than the disease incidence in Sidowayah District. Thus, disease incidence played an essential role in 100-seed weight and seed yield. Similarly, in a study conducted by Kuswantoro [20],100-seed weight is one of the main components affecting seed yield.
The absorbance ELISA value (AEV) results showed that all strains in the Sidowayah sub-district did not detect with SMV antiserum except W/Mlg 3288-7-9 line. Soybean lines (W/Mlg 3288-7-9, W/L.Temanggung-7-1, GK/L.Temanggung-7-9, GK/PI2000.485-7-2) and Anjasmoro variety in Girisubo District detected with SMV antiserum (Table 6). However, not all symptomatic plants reacted positively with antiserum against SMV. Because another virus, namely *Cowpea mild mottle virus* (CpMMV), can also cause mosaic symptoms in soybean plants [5].

4. Conclusion

The yield performance of F7 soybean lines resistant to *Soybean mosaic virus* in Girisubo District is lower than in Sidowayah District. Two lines have higher seed yields than Dering 1 and Anjasmoro varieties in both locations, namely W/Mlg 3288-7-9 and GK/PI200.485-7-17. The GK/PI200.485-7-17 line was the most superior soybean line because it had the highest number of branches, filled pods per plant, and 100-seed weight. In addition, there was the most consistent seed yield and resistance to SMV among all tested lines and control varieties in both districts.

5. References

[1] Pereira, L. , Trikoesoemaningtyas, and D. Wirnas. 2016. Genotype season interaction effects on the performance of soybean lines. Intern. Journal of Agronomy and Agriculture Research 9: 42-50.
[2] Carmello, V. and J.L.S. Neto. 2016. Rainfall variability and soybean yield in Paraná State, Southern Brazil. International Journal of Enviromental & Agriculture Research 2: 86-97.
[3] Andayanie, W.R., M. Lukito, and F. Chasanatun. 2021. Antifungal of cashew (*Anacardium occidentale* Linn) leaves, nut shells, and peduncle bagasse ashes extracts against sooty mould fungi (Capnodium sp). In: The 6th International Symposium on Applied Chemistry (ISAC) 2020. IOP Conf. Series: Materials Science and Engineering. 2021, 1011 (2021) 012039. pp. 1-6.
[4] Elisabeth, D.A.A. and Suhartina. 2019. Farmers' response on introduction of soybean cultivation technology under teak shade. In: International Conference on Biology and Applied Science (ICOBAS). AIP Conference Proceedings 2120, 040013. pp. 1-9.
[5] Andayanie, W.R. and N. Ermawati. 2019. Developmental effect of cashew nut shell extract against nymphal instar of silver leaf whitefly (*Bemisia tabaci* Genn.). In: The 2nd Int. Conference on Natural Resources and Life Science. IOP Conf.Proc.2019, 293. pp. 1-7.
[6] Andayanie, W.R., W. Nuriana, and N. Ermawaty. 2019. Bioactive compounds and their their antifeedant activity of cashew nut (*Anacardium occidentale* L.) shell extract against *Bemisia tabaci*, (Gennadius, 1889) (Hemiptera: Aleyrodidae**)**. Acta Agriculturae Slovenica 113: 281-288.
[7] Haling, R.E., R.J. Simpson, E. Delhaize, P.J. Hocking, and A.E. Richardso. 2010. Effect of lime on root growth, morphology and the rhizosheath of cereal seedlings growing in an acid soil. Plant and Soil 327:199-212.

[8] Boiteau, G., M. Singh, and J. Lavoie. 2009. Crop border and mineral sprays used in combination as physical control methods of aphid transmitted potato virus Y in potato. Journal Pest Management Science 65: 255-259.
[9] Hooks, C.R.R. and A. Fereres. 2006. Protecting crops from non-persistently aphid-transmitted viruses: A review on the use of barrier plants as a management tool. Virus Research 120: 1-16.
[10] Andayanie, W.R., N. Ermawaty, and R. Iswati. 2018. Use tillage system and botanical herbicide of cashew nut shell extract on losses nutrient and organic matter in the sloping land,". In: The 4th International Symposium on Applied Chemistry 2018, AIP Conference Proceedings 2024, Published. by the American Institute of Physics. 020018-1–020018-6.
[11] Koenig, R. 1981. Indirect ELISA methods for broad specificity detection of plant viruses. Journal of General Virology 55:53-62.
[12] Parvej, M.R., N.A. Slaton, L.C. Purcell, and T.L. Roberts. 2015. Potassium fertility effects yield components and seed potassium concentration of determinate and indeterminate soybean. Agron. J. 107:943-950.
[13] Roossinck, M.J. 2015. Plants, viruses and the environment: Ecology and mutualism. Review. Virology 47:271-277.
[14] Khanam, M., M.S. Islam, M.H. Ali, I.F. Chowdhury, and S.M. Masum. 2016. Performance of soybean under different levels of Phosphorus and Potasium. Bangladesh Agron. J. 19: 99-108.
[15] Gebrehana, Z.G. and L.A. Dagnaw.2020. Response of soybean to *Rhizobial* inoculation and starter N fertilizer on Nitisols of Assosa and Begi areas, Western Ethiopia. Environ Syst Res. 9:1-11.
[16] Yang, Q., G. Lin, L.V. Huiyong , C. Wang, Y. Yang, and H. Liao. 2021. Environmental and genetic regulation of plant height in soybean. BMC Plant Biology 21: 1-15.
[17] José da Silva, A., J.R.M. Filho, C.R.G. Sales, R.C.M. Pires, and E.C. Machado. 2018. Source-sink relationships in two soybean cultivars with indeterminate growth under water deficit. Bragantia, Campinas 77: 23-35.
[18] Kapoor, D., S. Bhardwaj, M. Landi, A. Sharma, M. Ramakrishnan, and A. Sharma. 2020. The Impact of drought in plant metabolism: How to exploit tolerance mechanisms to increase. Appl. Sci. 10:1-19.
[19] Diptaningsari, D. and A.A. Rivaie. 2021. Comparison study of growth and yield of three soybean varieties on acid upland soil of South Lampung. In: IOP Conf. Series: Earth and Environmental Science 648 (2021) 012066. pp. 1-7.
[20] Kuswantoro, H. 2016. Potential yield of acid-adaptive soybean promising lines in ultisols of Tanah Laut Regency, South Kalimantan Province, Indonesia. Biotropia 23: 52-57.

IOP Conf. Series: Earth and Environmental Science 1131 (2023) 012003 doi:10.1088/1755-1315/1131/1/012003

Is Long-term Cassava-based Agriculture Sustainable? (Case Study of Potassium Content in the Soil)

Ursulin Sacer Setyastika*, **Sri Rahayu Utami, Syahrul Kurniawan and Christanti Agustina**

Universitas Brawijaya, Malang, Indonesia.

*ursulinss@gmail.com

Abstract. Farmer's interest in planting cassava is relatively high because cassava is one of tolerant commodities toward drought, and low soil fertility. With minimum input, cassava still grow and produce. However, cassava absorb greater nutrients, especially potassium (K), leading to the depletion of essential nutrient availability. This research was carried out in Pati, Central Java, aimed to determine K nutrients status in the soil at the various durations of cassava land use. The research consisted of two factors, i.e 1) duration under cassava (five years, 10 yr, and 15 yr), and 2) slope (8 to 15% and 15 to 40%), each repeated four times. Soil and plant samples were taken at each location to determine K nutrients in the soil and plant (at harvest). Exchangeable K varies from low to moderate content, i.e. 0.44 to 0.50 cmol/kg (slope 1) and 0.22 cmol/kg (slope 2). Cassava land use showed an imbalance between input (fertilizer) and output (crop harvest) with negative results in the range of 114.59 to 524.78 kg K/ha. Exchangeable K content in the soils decreased with increasing K imbalance, meaning the longer period under cassava could lead to impoverishment of nutrients. Nutrient input should be increased to reduce the nutrient imbalance on the cassava farming system.

1. Introduction

Cassava (*Manihot esculenta*) is one of tolerant commodities toward drought, low soil fertility, and minimum input [1]. Cassava is able to produce in poor soils, even without adding organic fertilizer. This encourages farmers continuously cultivate cassava with minimum input. Cassava can be an alternative food ingredient besides rice and corn, where the government is actively promoting food diversification programs. Cassava removes more nutrients in soil, which can lead to nutrient depletion if planted with minimum input and without returning organic matter to the soil. Previous studies showed that cassava absorbs and consequently removed large amount of potassium, as well as nitrogen and phosphorus at lesser degree [1][2].

Cassava cultivation without balanced nutrient input is suspected as the cause of soil fertility degradation, because cassava absorbs greater nutrients which will impoverish the nutrient availability in the soil [2][3]. The longer period under cassava monoculture, the higher is the nutrient depletion. Since farmer tend to apply minimum fertilizer input, the imbalance nutrient will be probably occurred. [2] showed that continuous cassava cultivation with minimum input or without organic fertilizer, resulted in a lower content of organic C (0.50%), total N (0.02%), CEC (1.53 cmol/kg), K (0.02 to 0.06 cmol/kg), and Mg (0.04 cmol/kg) in the soils in comparison to forest, rubber, cashew, and sugarcane land-use [3]. This will lead to degraded soil physical, chemical, and biological properties. Therefore, this research was attempted to study the effect of long-term period under cassava cropping systems on nutrient status (especially K) in the soil.

2. Materials and methods

2.1. Material

The research was carried out in Gembong District, Pati, Central Java. Gembong District is one of nine sub-districts in Pati which produce cassava. Gembong District has the highest potential for cassava productivity [5]. Almost every land in Gembong District is planted cassava due to the adequate benefits of cassava, the development of cassava processing industry, and high tolerance in poor soils [4]. The rainfall is about 196 mm (over the last five years) and the type of parent material is volcanic material from Muria volcanoes and generally classified as Inceptisols.

2.2. Methods

We selected sampling plot location, having different slope and cultivation period under cassava, but similar in soil type, cassava variety, and cultivation practice. Cassava commodities is distributed in slope 1 (8 to 15%) and slope 2 (15 to 40%) in Gembong District [6]. Hence, Semirejo Village was selected as the sampling location in slope 1 (longitude of 110°59.711'E and latitude of 06°42.729'S) and Bageng Village as the sampling location in slope 2 (longitude of 110°56.792'E and latitude of 06°41.355'S).

The research was conducted on three plots: (1) Cassava monoculture for five years, (2) Cassava monoculture for 10 yr, and (3) Cassava monoculture for 15 yr, at two different slopes: slope 1 (8 to 15%) and slope 2 (15 to 40%). Each treatment was repeated four times, totally 24 research plots. Each plot was delineated 20 m x 20 m in size, in which we sampled five points in diagonal positions. Soil samples was collected at two soil depths (0 to 20 cm, and 20 to 40 cm). Plant was also sampled from every plot. Exchangeable K in the soil and total K adsorbed were analysed in laboratory.

Nutrient balance is calculated from the difference between input and output of the agricultural system. In this research, nutrient balance was calculated based on the fertilizer applied by farmer (input) and harvest (output). The fertilizers used by cassava farmers in Gembong District are 300 kg/ha Urea, 207.5 kg/ha Phonska, and 100 kg/ha ZA. The collected data was statistically analyzed using univariate test, followed by Tukey test at 5% level using SPSS.

3. Results and discussion

3.1. Impact of long-term cassava on potassium content in soil

The results showed that slope significantly ($P<0.05$) affected K content in the soil. Table 1 indicated that soils at slope 1 (8 to 15%) contained significantly higher K than soils at slope 2 (15 to 40%). This result could be associated to a larger K loss by surface runoff, which possibly occurs in larger degree at slope 2 than slope 1. Whereas, there was unsignificant difference between K nutrients in the cassava monoculture for five year, 10 yr, and 15 yr, both at depth 0 to 20 cm ($P=0.93$) and 20 to 40 cm ($P=0.33$). This indicated that K loss through erosion might be larger than K removal by cassava. [2] showed that cassava cultivation caused more soil loss because of erosion process in comparison with other crops (peanut, rice, cotton, soybean, potato, maize, and sweet potato). This is because of wide spacing used in cassava cultivation and growing cassava in slope which can cause serious gully or sheet erosion. Other study [7] also showed that nutrients (N, P, K, etc) contents were decreased if cassava was planted in wider spacing and at early rainy season, leading to a more open canopy and an extensive erosion. Consequently, the topsoil which originally containing higher organic matter will be removed and transported to lower slope. In addition, soils at slope 2 had smaller CEC than slope 1, indicating lower capability to retain cations, and hence susceptible to cations removal by erosion. A high soil CEC would able to absorb and provide nutrients better than soils with a low CEC [8]. [9] showed that soil with low CEC, normally due to a low clay content and organic matter, experienced larger leaching of nutrients (NO_3^-, NH_4^+, K^+, and Mg^{2+}).

Table 1. Exchangeable K content in the study soils

Treatments	Exchangeable K (cmol/kg)	
	at depth 0 to 20 cm	at depth 20 to 40 cm
Duration under cassava land-use (year):		
P1 (cassava monoculture for five years)	0.38	0.44
P2 (cassava monoculture for 10 yr)	0.36	0.40
P3 (cassava monoculture for 15 yr)	0.38	0.38
Tukey 5%	ns	ns
Slope (%):		
L1 (8 to 15%)	0.44 a	0.50 a
L2 (15 to 40%)	0.22 b	0.22 b
Tukey 5%	0.08	0.07

Note: The same letter notation in same column each factor showed no significant difference in Tukey test 5% level.ns = no significant.

Exchangeable K content in the study soils was classified as low to moderate with range of 0.20 to 0.38 cmol/kg (at depth 0 to 20 cm) and 0.23 to 0.44 cmol/kg (at depth 20 to 40 cm). The results indicated that higher exchangeable K occurred in the soil under five years cassava monoculture, both at 0 to 20 cm and 20 to 40 cm depths. Longer period under cassava cultivation potentially removed K at larger amount than shorter period. This could be also due to higher organic C in five years than 10 yr and 15 yr cassava cultivation. Soils in five years cassava cultivation contained 0.77% organic C (at depth 0 to 20 cm) and 0.60% organic C (at depth 20 to 40 cm), higher than 0.56 to 0.62% organic C (at depth 0 to 20 cm) and 0.46 to 0.49% organic C (at depth 20 to 40 cm) in 10 yr, 15 yr cassava cultivation. Soils having higher organic matter, then the nutrients availability was also higher. This is because of soils with higher organic matter content will have higher soil CEC and nutrient retention capacity, hence reducing nutrients leaching. [10] showed that the K content in the soil of cassava monoculture decreased along with the long-term used of cassava cultivation, in which cassava monoculture for four and 10 yr had lower K content than cassava monoculture for one year.

3.2. Effect of potassium content in soil and potassium balance

Nutrient balance is a basic concept for understanding nutrient requirements in soil and plants. Nutrient balance was calculated from the difference between input giving to the system and output coming out of the system [11]. Nutrient output was generally higher than nutrient input in long-term cassava cultivation (Figure 1). This indicated imbalance nutrients in the study cassava land use system. Continuous application of potassium fertilizer (207.5 kg/ha Phonska) did not cope the K removal by cassava yield.

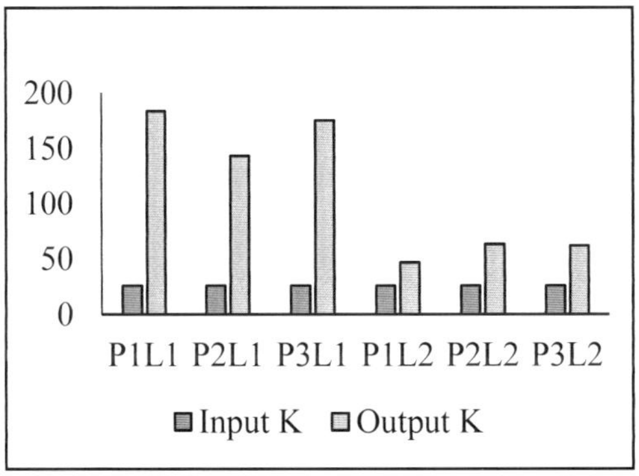

Figure 1. Input and output on cassava land-use

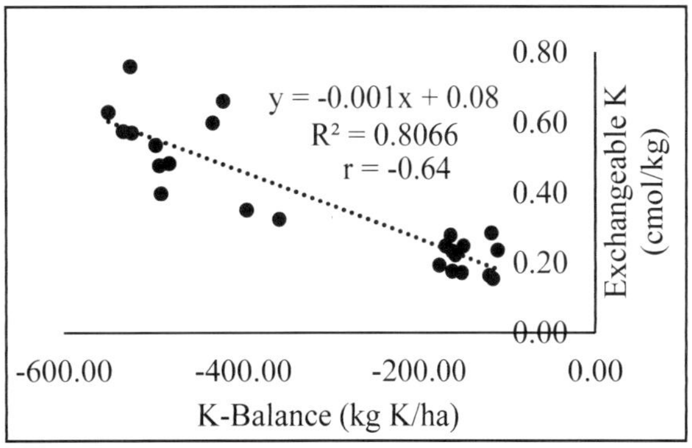

Figure 2. Relationship between exchangeable K-content in the soil and K-Balance

The results indicated that K balance in cassava cultivation will be negative, if farmers only add inorganic fertilizers or imbalance fertilization. Figure 2 showed that there is an imbalance of K nutrient which is indicated by a negative value. K imbalance is strongly and negatively correlated with K content in the soils (r=-0.64). The regression equation was Y = -0.001x + 0.08, where Y was K content in the soil and X was the K imbalance. A larger K imbalance would result in a larger decrease of K content in the soil, reflecting the impoverisment of soil fertility. As cassava cultivation continuos with time, the longer period under cassava cultivation, the larger imbalanced use of K, the lower is the fertililty status of the soils. K imbalance had affected about 80.66% on K content in the soils, and about 19.34% was other factors. Cassava needs K nutrients in large quantities (187 kg K/ha) for accumulating carbohydrates and increasing cassava production [12]. [7] resulted that there was an K imbalance in cassava monoculture (-87 kg/ha), because there was no K fertilizer which given to the soil. Meanwhile, K imbalance decreased to -57 kg/ha due to the application of K fertilizer about 30 kg K_2O/ha. This means that nutrients content in the soil has an important role in nutrient balance. By adding fertilizer in the soil will decrease the nutrient imbalance, because adding fertilizer will increase the K content in the soil too. Furthermore, the addition of organic matter, or applied the balance fertilization (inorganic mixed with organic fertilizer) could increase K content and K balance in the soil.

4. Conclusions

The duration under cassava land-use unsignificant affected K content in the soil. However, there was tendency that the longer the period under cassava cultivation, the larger imbalanced K between K removal by yield and K input by fertilizer. This will lead to larger decrease of exchangeable K content in the soils. Continuous application of inorganic fertilizer (300 kg/ha Urea, 207.5 kg/ha Phonska, and 100 kg/ha ZA) did not reduce K imbalance on cassava land-use. Exchangeable K content in the soils strongly related to the K imbalance use in cassava cultivation. Exchangeable K content in the soil at slope 1 was higher (0.44 to 0.50 cmol/kg) than K content in the soil at slope 2 (0.22 cmol/kg), which could be related to higher erosion. This study showed that management should be improved to reduce the imbalanced K in cassava-based cultivation, especially when cultivated on sloping area.

5. References

[1] Pongsivapai, P., C. Thongjoo, J. Romkaew, and T. Inboonchuay. 2016. Effect of Fertilizer Management in Combination with Soil Conditioner on Yield of Cassava Cultivated on Coarse-Textured Soil in Thailand. *Modern Applied Science* 10 (11): 239-247.

[2] Howeler, R.H. 2014. *Sustainable Soil and Crop Management of Cassava in Asia.* Centro Internacional de Agricultura Tropical (CIAT). Cali, Colombia.

[3] Ispandi, A. 2002. Peningkatan Produksi Ubi Kayu di Lahan Kering Iklim Kering. *Buletin Palawija* 3: 17-25.

[4] Agustina, C. 2015. Menengok Sentra Ubi Kayu di Pati. *Available online at* https://tabloidsinartani.com/detail/indeks/nusantara/2787-menengok-sentra-ubi-kayu-di-pati .

[5] Nugraha, H.D., A. Suryanto, and A. Nugroho. 2015. Kajian Potensi Produktivitas Ubi kayu (*Manihot esculenta* Crant.) di Kabupaten Pati. *Jurnal Produksi Tanaman* 3 (8): 673-682.

[6] Badan Penyuluhan Pertanian Kecamatan Gembong (BPP Gembong). 2019. Programa Penyuluhan Pertanian Kecamatan Gembong Tahun 2020. Balai Penyuluhan Kecamatan Gembong, Kabupaten Pati. Pati.

[7] Wijanarko, A., and B.H. Purwanto. 2018. Effect of Long of Landuse and Cropping System on Soil Fertility and Cassava Yield. *Journal of Degraded and Mining Lands Management* 5 (4): 1327-1334.

[8] Hardjowigeno, S. 1987. Ilmu Tanah. 1st ed. Mediyatama Sarana Perkasa. Jakarta.

[9] Handayanto, E., N. Muddarisna, and A. Fiqri. 2017. Pengelolaan Kesuburan Tanah. Universitas Brawijaya Press (UB Press). Malang.

[10] Saleh, N., A. Taufiq, Y. Widodo, and T. Sundari. 2016. Pedoman Budidaya Ubi Kayu di Indonesia. IAARD Press. Jakarta.

[11] Husnain, A. Kasno, and S. Rochayati. 2016. Pengelolaan Hara dan Teknologi Pemupukan Mendukung Swasembada Pangan di Indonesia. *Jurnal Sumberdaya Lahan* 10 (1): 25-36.

[12] Subandi. 2011. Pengelolaan Hara Kalium untuk Ubi Kayu pada Lahan Kering Masam. *Buletin Palawija* 22: 86-95.

IOP Conf. Series: Earth and Environmental Science 1131 (2023) 012004 doi:10.1088/1755-1315/1131/1/012004

Urban Agriculture Exploration Strategy: Case Study of Mushroom and Freshwater Fish Cultivation in West Surabaya

S Sutini*, G Guniarti, W Widiwurjani, Nora Augustien, Didik Utomo Pribadi

Agrotechnology Department of Agriculture Faculty UPN "Veteran" Jawa Timur, Indonesia

*tien.basuki@gmail.com

Abstract. Urban agriculture is constrained by the increasingly unstoppable level of urbanization which has an impact on the agricultural land in urban areas reduction. This kind of problem need to find alternative solutions, they are; integrating urban agriculture with the environment, partnership activities, e.g., with universities and sustainable diversification. The purpose of this article is to explore urban agriculture: a case study of mushroom and freshwater fish cultivation in West Surabaya. The writing of this article includes case studies related to oyster mushroom cultivation, inland waters and their diversification. Data collection methods include intensive documentation and interviews with farmers, city governments, partnerships and consumer organizations. The data collected was analyzed by matching the data with urban agricultural scheme as a literature study result. The results obtained show that there are partnerships with universities as evidenced by the existence of Community Service Program activities carried out together with urban farmers. And also the sustainability of processed fish and mushroom diversification. However, it is necessary to have a processed diversification training program as well as bringing together prospective consumers more intensively, and alternative waste handling that is safe for the environment.

1. Introduction

Urban agriculture is interpreted as agricultural activities both within the city and suburbs for cultivating, maintaining, harvesting, processing, distributing various agricultural products both food and non-food by maximizing human resources, materials, products and services in urban areas [1]. Urban agriculture is also an efficient movement, establishing people's lifestyles, sustainable urban planning design, utilization of green open space, utilization of local food, and related to the green land existence [2]. The area of agricultural land in urban areas is very diverse, this is due to the competition for land use between the agricultural sector and other sectors outside of agriculture. The area of agricultural land in urban areas can be referred to as the area of the yard, it should be cultivated for agriculture up to 30% of the yard area, offices and public facilities [3].

Data from the Central Statistics Agency projects that population migration from rural to urban areas in 2020 will increase to 56.7% and in 2025 it will increase to 60%, which must be resolved [4]. Mean while, according to FAO, urban agriculture has three criteria, they are: 1. as one of the origins of food supply and

a household-scale food security system; 2. positive activities that produce green open spaces and utilization of household waste; 3. increase income and provide jobs for urban communities [5].
The implementation of urban agriculture in several developed countries such as in America is used for community outreach activities, individual capacity building and technical capacity building. The implementation of urban agriculture in Beijing, China, serves to produce food and diversification of organic agriculture. The implementation of urban agriculture in Europe, precisely in the city of Barcelona, is carried out by elderly communities who produce vegetables which have positive implications for the needs of home-cooked vegetables. The implementation of urban agriculture in Russia that is in the city of Petersburg produces a wide variety of vegetables grown in greenhouses [6]. Urban agriculture can be part of the solution to meet food needs, as well as prospecting for environmentally sound entrepreneurship. Urban agriculture can be an alternative to help food security and can reduce poverty in urban areas. The implementation of urban agriculture in the villlage of Made, West Surabaya has succeeded in developing chili commodities which can increase farmers' income and for self-development of the community. The purpose of this article is to explore urban agriculture: a case study of mushroom and freshwater fish cultivation in West Surabaya.

2. Methodology

2.1. Methodological Approach

The methodological approach applies a case study centered on a phenomenon that occurs in the field to be able to describe a case. The description of a case is obtained by exploring the case in detail with observation [7].

2.2. Documentation and Interview

Documentation, and spontaneous and open interviews with: 1. farmers with intensively free structures, 2. municipal governments, 3. partnerships, 4. consumer organizations [8]. The repeated interviews were conducted five times, the interviews are in Table 1.

Table 1. Interview

Category of interviewee	Opinions of people regarding urban agriculture
Urban farmers	1. Oyster mushroom farmer 2. Freshwater fish/ shrimp farmers
Urban agriculture association	3. “Makmur” farmers Association 4. “Wiji” farmers Association
Village officers	5. Secretary of the village 6. Village treasurer
Head of Village	7. Head of Village
District officers	8. District administration officer
Subdistrict head	9. Subsdistrict head
Home industry	10. Home manufacturers
Consumers	11. Urban agricultural consumers

2.3. Subjects of the Study

The subjects of this study were 2 farmers of oyster mushroom and inland waters, the village head and his 2 officers, the sub-district head and his 1 officer, 2 partners, with a duration of six weeks

2.4. Data Collection Technique

Data collection techniques applied are: interviews, observation, documentation. First, the interview was conducted with a "how and why" question on the subject to understand the causes and effects of a problem. Second, observations were made to obtain information and to make notes about the observed subjects. Third, documentation in the form of photographs, videos and written materials [9]

2.5. Data Analysis
Data analysis was carried out in three stages. The first stage is reducing data by simplifying, and revising the data that appears while in the field. The second stage is organizing the data in the form of tables or pictures to draw a conclusion. The third stage is drawing conclusions and verification which is carried out continuously during the analysis [10].

3. Results and Discussion

3.1. Results of the survey
The results of observations - surveys showed that in the Balas Klumprik village, precisely in the 4th hamlet,
the potential for urban agricultural products is in the form of freshwater fish such as shrimp, catfish, nile tilapia and oyster mushrooms. Observations were made to explore the potential of natural resources as a basis for further activities. The survey results are presented in Figure 1

Figure 1. The potential of urban agricultural products is in the form of freshwater fish such as shrimp, catfish, nile tilapia and oyster mushrooms in Balas Klumprik Village, Wiyung sub-district, West Surabaya.

3.2. Results of documentation and interviews
The results of documentation and interviews were research tools that can be carried out very effectively and flexibly with free/unbound question sentences in order to be able to explore in-depth data from the participants.

3.3. Partnerships
The partnership from this case study can be seen from the partnership between farmers and the local community with the University, farmers with some consumers who are in the area around the Balas Klumprik village in west Surabaya. Partnership with the university was established in the form of the students from University of Pembangunan Nasional East Java that had being permitted to conduct Student Community Service (KKN) with additional educational activities for elementary school children & the diversification process of processed agricultural productsin the village [11]. Partnerships with institutions/industry and university are important because urban farmers are very vulnerable at agricultural technology, besides that they are also vulnerable at finance for initial work of agricultural. Relevant to the opinion [12] that the population in Indonesia is compared to experts and technology, Indonesia is still far behind with other countries which has implications for the agricultural system in urban areas. Diversification process of urban agricultural products aside from enriching food diversity, it is also a way to preserve materials. This preservation is intended for when harvesting exceeds capacity, it is hoped that agricultural products will not be damaged quickly. Diversification includes business management result of oyster mushroom harvests according to the research [13]

3.4. Results of interviews

Interviews in case study research with an independent structure were applied to explore the opinions of office holders, they were the Village head and village officials as well as Sub-district officials, urban agricultural farmers and urban agricultural organizations and consumers. The first interview addressed to the Village Head including the following: "What is the opinion of the Village Head regarding to the existence of urban agriculture in the area of the village he leads, can it meet the needs of urban farmers?" In this interview, information was obtained that broad urban agriculture still needs to be improved, but for small families it can at least help the family's own needs, besides that the quality of food commodities is also getting healthier, the environment is more friendly and the oxygen supply is more abundant and productive [14]. The second interview was asked to the village officials: "What do you think about how is the effect of urban agriculture on the cleanliness of the environment around urban agriculture and does the land that used have legal status?" In this second interview, the answer was that urban agriculture related to freshwater fish aquaculture that the waste disposal can still be tolerated because the surrounding area is still a vacant land and the flow of waterways is still controlled in volume. However, this vacant land is partially owned by the government. Due to the illegal status of the land, some farmers were evicted because the land they occupied would be built by the government for public facilities. According to the research [15] it is stated that because farmers do not have land, so they use vacant land illegally. The third interview was delivered to the local Sub-district, it was: "Why does urban agriculture can increase community empowerment, even though the land is limited?." This third interview obtained information that limited land is a challenge for creative residents, for example those with limited land but the community uses a vertical system and grows plants that are adapted to the land area, or yard. It was also investigated by the researchers [16] that people used literacy for their daily agricultural activities. The next interview was given to urban farmers, it was: "How do you grow mushrooms in narrow urban areas and a less supportive environment?" The information obtained is that they were active in farmer group associations that have been fostered by the relevant agencies. This activity of urban farmers is in accordance with research [17] that farmers applied intensification.

4. Conclusion

The conclusion of this article is that if urban agriculture being managed properly and carefully, it will help improve people's nutrition, and if it is also carefully driven by the community and government, urban agriculture will improve the environment. The environment becomes beautiful, green and clean while supporting the reduction of carbon dioxide gas. If carried out on a land area that is supported by public facilities, urban agriculture will be able to produce and can be used as a recreational place for the surrounding community. If the sewerage can be arranged well, urban agriculture based on inland waters will also become a means of recreation that can also improve the nutrition and economy of the community.

References

[1] Smith J, Ratta A, and Nasr J 1996. Urban Agriculture: Food, Jobs and Sustainable Cities. PublicationSeries for Habitat II. vol I. New York: United Nations Development Program (UNDP)
[2] André B K. and Viljoen A 2014. Second Nature Urban Agriculture- Designing productive cities Routledge - Taylor & Francis Group.
[3] Sampeliling S, Santun R P S, Nurisyah S, Pramudya B 2012. Sustainable Urban Agricultur Development Policy: A Case Study in Jakarta. Lembaga Penelitian dan Pengabdian Masyaraka IPB, Gd. Andi Hakim Nasution Lt.3, Kampus IPB Darmaga, Bogor
[4]. Indahri Y 2017. Population Management Challenges in The City of Surabaya. Pusat Penelitian Badan Keahlian DPR RI
[5]. FAO. 2003. Trade Reform and Food Security – Conceptualizing The Linkages. FAO.
[6]. Ahmad R F, Annisa N I, Agustin H 2016. Urban Agricuture : Urgency, Role, and Best Practice Jurnal Agroteknologi, vol. 10 no. 01
[7] Raco J R, 2010. Qualitative research methods. Jakarta: PT. Grasindo, 49-50. S.L.

[8] Jackson, 2008. Research methods. USA: Thomson Wadsworth, 88.
[9] Moleong J L 2007. Qualitative research methods. Bandung: PT. Remaja Rosda Karya, 219.
[10] Daniel A, 2009. Perception gaps between headquarters and subsidiary managers. Germany:Gabler, 138.
[11] Sutini, Guniarti, Abubakar R W 2016. Kuliah Kerja Nyata – Learning of Society Empowerment Base of Entrepreneurship Culture in Kelurahan Balas Klumprik SubDistic Wiyung - Surabaya. Ciputra University Surabaya, Indonesia
[12] Bagyo Y. Moeliodihardjo, Biemo W S, Satryo S B, Hatakenaka S 2013. Strategy Development of Higher Education, Industry, and Government Partnerships in Indonesia. Partnership Program for Capacity Development and Education Analysis (ACDP)
[13] Muhammad I W, Tripama B, Prayuginingsih, Taufik T W 2021. Product Diversification of Oyster Mushroom to Support Community Economy in Jember District. Jurnal Ilmiah Pengabdian kepada Masyarakat, 7 (1) 13–21
[14] Ayu K G, 2013. The role of urban farming for attracting consumer Interest to buy residential real estate in surabaya. Master's Program in Real Estate Planning Expertise, Department of Architecture, Faculty of Civil Engineering and Planning, Sepuluh Nopember Institute of Technology, Surabaya
[15] Sadyohutomo M , 2018. The benefits of an agrarian reform model in Indonesia IOP Conf. Ser.: Earth Environ. Sci. 202 012030
[16] Rahmawati D, SulistyarsoH, Ariastita P G, Paramasatya D A, Yusuf M, 2018. Smart kampung for Surabaya smart city: Criteria redefined. IOP Conf. Series: Earth and Environmental Science 202
[17] Syaifudin A and Carsjens G J, 2018. Exploring farming strategies in a metropolitan area: case study of inland aquaculture in Bogor Regency IOP Conf. Ser.: Earth Environ. Sci. 202 012026

Growth of *Arengapinnata* Seedlings on Three Different Media

Budi Prakoso*, R Widarawati

University of Jenderal Soedirman, dr. Soeparno street, Karangwangkal, Kabupaten Banyumas 53123, Indonesia

*prabud2001@yahoo.com

Abstract. Aren (*Arengapinnata*) trees are the second most important palm trees in Indonesia. However, little information on the effect of media on growth of aren (sugar palm) tree seedlings. The growth of six monts old seedlings on three media, namely soil, soil+Trichocompost, and sawdust+Trichocompost was studied. It was found that the growth of seedlings 45 days after replanting on soil+Trichocompost was higher than that of on sawdust+Trichocompost; However, the growth of the seedlings on soil media was not significantly lower than that of on soil+Trichocompost, nor significantly higher than that of on sawdust+Trichocompost.It was recommended that soil+Trichocompost and sawdust+Trichocompost can be used for aren seedling growing media.

1. Introduction

Aren (*Arengapinnata*) is one of important smallholder industrial palm trees in Indonesia. However, most of the trees donot been planted from seeds. The trees grow naturally on household field, on forests and on other marginal land sites [1].Most farmers replanted seedlings that they found grown naturally on forests or others siteson their backyards or on their own lands [2]. The aren trees are the second most important palm trees in term of the number of plants, distribution areas and economic values. The most economic share for the farmer is brown sugar made from sap of flower stems. Price ofarenbrown sugar is higher than that of coconut brown sugar or cane white sugar and farmers can get 1-2 kg sugar per plant per day. The sap can also be fermented for making ethanol or beverage. Processed endosperms of young fruits of aren are eaten as cocktail, local refreshment, and medicine for curing many diseases. Flour of aren stems is natural edible food, and the leaf sheath is a source of a tough, black fibre [3].

Most farmers did not cultivate the arentrees intensively.The trees grew naturally, without fertilizer and pesticide application[4]. Most farmers did not germinate the aren seeds on nursery beds since germination takes 1- 2 months and 8-10 months for the seedling can be replanted on polybags or even 2-3 years for seedlings ready to be planted in fields [5].

Little information on goodmedia for germinating and raising seedlings of aren. Good growing media should support plant growth, holds water and nutrients, good drainage, and should not easily degraded [6]. Soil was one of media commonly used in nursery, however it heavy. Organic materials commonly were added to lighter and improved the quality of soil media. Some of the organic materials were compost, cocopeat, charcoal, and sawdust [7]. Compost is degraded organic materials that turns into humus-like materials. It is produced in household and industrial sites. It contains essential macro and micro nutrients needed by plants. In some cases it contains heavy metals and harmful products. However, nutrient content of compost varied [8].SNI number 19-7030-2004 is Indonesian standard of domestic compost. It should contain at least 0.4% Nitrogen, 0.1% P_2O_5, 0.2% K_2O, and C/N ratio in the range of 10-20 [9]. Commercial organic fertilizers based on Ministry of Agriculture decree number 261/KPTS/SR.310/M/4/2019 should contain at least 2% of Nitrogen+P_2O_5+K_2O and C/N ratio 25 [10]. Addition of organic fertilizer increased number of leaves, but it did not increased the height of seedlings of aren [11]

Sawdust was the cheapest organic material, since sawdust is one of industrial sawmill waste that dumped on sawmill sites and it becomes breeding site of *Orictesrenoceros,* the pest of coconut and others palm trees. Some finding on the effect of sawdust based media on germination, growth and yield of crops varied. Growth and yield of shallotgrown on sawdust were lower than that of on rice husk charcoal and on cocopeat [12]. Other study found that vegetative characters of papaya seedlings grown on soil+sawdust (1:1) was the lowest compare to that of on soil+compost (1:1); soil+compost+ husk charcoal (1:1:1); soil+compost+cocopeat (1:1:1); or soil+compost+Albasia sawdust (1:1:1) [7]. High soluble tannin in the sawdusk might retard the growth of plants [5]. Soaking sawdust with water and pouring the water every day for removing the soluble tannin for 7 days resulted in that height of seedlings of pepper grown on sawdust that has been soaked for 7 days was better than that of grown on soil [13].

This research studied the growth of sugar palm seedlings on three different media, namely soil, soil + trichocompost(1:1 v/v), and sawdust + Trichocompost(1:1 v/v). Substitution of soil with trichocompost and sawdust reduced the weight of the media. It will reduce cost for transporting the seedlings.

2. Materials and methods

The experiment was carried out in polybags, atSokaraja Tengah village, Sokaraja, Banyumas. It was carried out from April until June 2021. Six monts old of seedlings on 10 cm diameter polybags were bought from a local farmer from Sunyalangu village, Banyumas, Central Java, Indonesia. The seedlings were replanted on 20 cm diameter polybags filled with media according to the treatments.One seedling in each polybag.

Anonfactorial experiment with sixteenth replications was arranged in Randomized Complete Block Design. The treatment was types of media i.e. soil, soil+Trichocompost (1:1 v/v), and sawdust+Trichocompost (1:1 v/v). The thichocompost contains 1.8% Nitrogen, 1.04 % P_2O5, 1.3 % K_2O and 175 organic C. The trichocompost has C/N ratio of 10-15. Observed variables were plant height and height of new buds. ANOVA was used for analyzing data followed by Least Significantly Different (LSD) if F test was significant

3. Results and discussion

Table 1.plant height of seedlings grown on different media

Media	Plant height (cm)		
	1 st observation	2nd observation	3rd observation
sawdust+trichocompost	36.7a	48.1a	50.6a
Soil	35.7a	45.3a	55.1ab
soil+trichocompost	37.6a	41.5a	62.0b
LSD 0.05	4.83	7.19	8.68

Table 1.showed that the height of seedlings of aren grown of all media in the first and second observation was not significantly different. However in the third observation the height of seedlings of

aren grown on soil+Trichocompost was heigher than that of on sawdust+Trichocompost, but it was not significantly different to the height of aren seedlings grown on soil. This finding supported the finding that addition of organic fertilizer did not improved the growth of seedling of aren [11] The height of aren seedlings grown on sawdust+Trichocompost was not significantly lower than that of on soil+trichocompost. Figure 1.showed that the height of aren seedlings has a linear relationship to the time of observation.

Average height of new bud of seedling on all media on the third observation did not significantly different. The average height of new bud of seedlings grown of sawdust+Trichocompost, soil, and Soil+Trichocompost was 18.03 cm, 18.62 cm. and 15.77 cm respectively. This finding supported the finding that sawdust could be used for media of chilli [12] and shallot [13] but contrary to the finding that papaya seedling grown on sawdust+compost+soil without nutrient solution addition was the lowest [7].

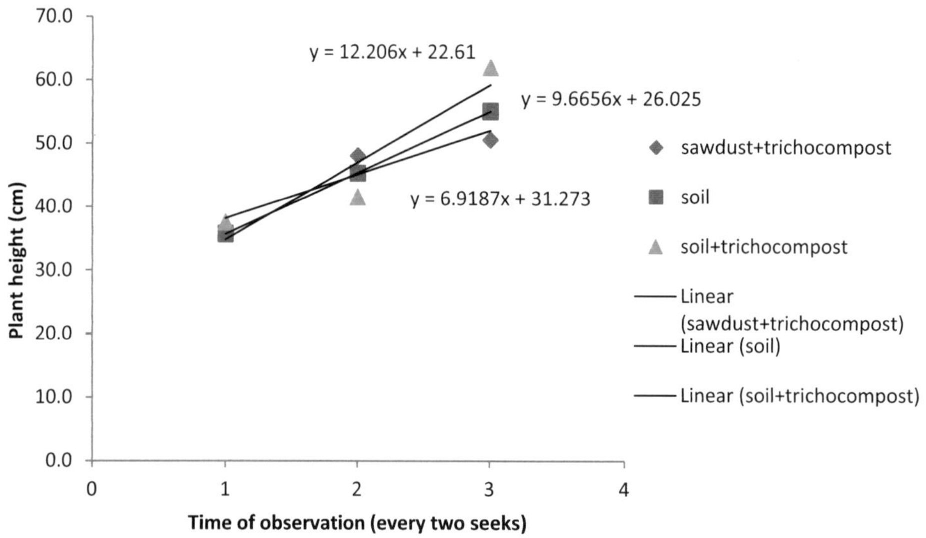

Figure 1. Effect of media on plant height

4. Conclusions

Substitution of 50% of soil with trichocompost improved the growth of seedlings. There is possibility that media consisted of soil+trichocompost is the best media for growing seedling of aren plants, The growth of seedlings on saw dust+trichocompost media waslower than that of in soil+trichocompost, but the growth of seedlings on saw dust+trichocompost media was similar tothat of on soil

References

[1] TropicalPlants Database, Ken Fern. tropical.theferns.info. 2021-11-03. <tropical.theferns.info/viewtropical.php?id=Arenga+pinnata>

[2] Widarawati, R., D. Prajitno dan P. Yudono. 2008. Kajianbudidayaaren (*Arengapinnata*Merr) dalam sistemagroforestri di Wilayah Jawa Twngah dan DIY. *Seminar Nasional Agroforestri.* Fak.Pertanian UNS – INAFE 2008, Solo.

[3] DirektoratTanamanTahunandan Penyegar, 2019*Mengenal Benih Aren Yang Baik.* Online. http://tanhun.ditjenbun.pertanian.go.id/web/page/title/3233/mengenali-benih-aren-yang-baik?post_type=informasi. Diakses 5 Juli 2021

[4] Rofik, A and E. Murniati. 2008. Pengaruh perlakuan deoperkulasi benih dan media perkecambahan untuk meningkatkan viabilitas benih aren (Arenga pinnata (Wurmb.) Merr.). Jurnal Agronomi Indonesia (Indonesian Journal of Agronomy), 36(1): 33-40

[5] Bernhard M R 2007. Teknik budidaya dan rehabilitasi tanaman aren. Buletin Palma, 33: 67-77.

[6] Bachtiar S., M. Rijaland D. Safitri. 2017.Potensispons sebagai media alternatifbudidayasayurandengansistemhidroponik.J. Biol. Sci. Edu.. 6: 52–60

[7] Prajwalita D T, S. R. Suparto and B. Prakoso. 2018.Growth of papaya cv. Callina seedlings on four types of planting media supplemented with different doses of AB-MIX nutrient solution. IOP Conf. Series: Earth and Env. Sci. 250 012029

[8] Mladenov M. 2018. Chemical composition of different types of compost. J. Chem. Tech. Metal., 53(4): 712-716

[9] Ngapiyatun S,A. Rahman, H. Aziza,B. Winarni, Wartomo. 2020.Pemanfaatanlimbahsampahkotasebagaikompos. Buletin LOUPE 16(2): 1-6

[10] KementerianPertanianRepublikIndonesia. 2019.Persyaratanteknis minimal pupukorganil, pupukhayati, danpembenahtanah.http://simpel1.pertanian.go.id/api/dokumen/regulasi/dokumen-1579833905542.pdf

[11] Prayoga F, R. S. Budi, F. M.Simbolon.2020.Pengaruhpemberianpupukorganikdan air kelapaterhadappertumbuhanbibittanamanaren (ArengapinnataMerr). AGRILAND JurnalIlmuPertanian 8(1):79-83

[12] Langgeng R. H., E. W. Tiniand B. Prakoso.2019.Pertumbuhan bibit cabai pada media serbuk gergaji kayu sengon dengan perendaman air.Agroteh. Res. J. 3: 97–102

[13] Sugiyanta D A,E. Rokhminarsi and B. Prakoso. 2021. Growth and yield of shallot (*Allium ascalonicum* L.) on different types of media and nutrient solution on hydroponic wick system IOP Conf. Ser.: Earth Environ. Sci. 653 012061

IOP Conf. Series: Earth and Environmental Science 1131 (2023) 012006 doi:10.1088/1755-1315/1131/1/012006

Socio Characteristics of Citronella Farmers in Kedungrandu Village, Banyumas District

Lutfi Zulkifli*, Anny Hartati, Tatang Widjojoko, Irene Kartika Eka Wijayanti, Adwi Herry Koesoema Elyanto, Alpha Nadeira Mandamdari, Rifki Andi Novia, Sunendar

Jenderal Soedirman University, Purwokerto, Indonesia

*lutfizulkifli@unsoed.ac.id

Abstract. Citronella is one of the leading commodities that have competitiveness and high export demand. Citronella is one of the primary ingredients for making essential oils. However, there are not many studies on the individual characteristics of Citronella farmers from a socio-economic point of view. This study aims to understand and analyze the personal characteristics of Citronella farmers in terms of gender, age, last education, farmers' primary and secondary jobs, experience in farming and cultivation of Citronella, training attended by farmers, and the number of dependents of each farmer's family. The research method used a survey method with the Simple Random Sampling technique. The data analysis tool used is descriptive statistics analysis. The results showed that the majority of Citronella farmers are male, aged 41-60 years. Most citronella farmers are educated up to elementary school and have their primary job as farm laborers. Citronella farmers mostly have side jobs as farm laborers for other commodities and, on average, have more than 20 years of experience in farming. However, the majority of citronella farmers only have less than three years of experience cultivating citronella.

1. Introduction

Citronella is one type of plant used to make essential oils that have developed [1]. This commodity is a resource of foreign exchange for increasing state income through its exports to various countries. The results of the distillation of fragrant citronella leaves will produce an essential oil called Citronella Oil. In the international market, this fragrant citronella oil knows as "Citronella Oil of Java." Citronella Oil is used as various basic raw materials, including perfumes, aromatherapy oils, medicines, or cosmetics. Based on the five largest essential oil-importing countries from Indonesia, export value is constantly increasing. Theimportance of crucial oil exports to the United States in 2020 contributed to the most considerable export value of 126,677$ [2]. This potential is an excellent opportunity for Indonesia to increase the production of essential oil-producing commodities, one of which is citronella. One of the districts in Central Java that has a high potential to develop citronella oil is Banyumas Regency. Banyumas Regency is a potential area for the cultivation and development of citronella oil. In this district, there is an essential oil processing plant capable of exporting 3,000 tons per year. Generally, citronella in this place is grown on plantation lands, both on people's plantations and government property. One of the plantation lands planted with citronella plants is in KedungranduVillage, Patikraja District. The purpose of this study was to identify the socio-economic characteristics of citronella farmers in Kedungrandu. The features of farmers [3], as the primary control holder in farming, describe how their encouragement, characteristics, self-concept, values, knowledge, experience, and expertise in carrying out their farming.

IOP Conf. Series: Earth and Environmental Science 1131 (2023) 012006 doi:10.1088/1755-1315/1131/1/012006

Aspects of citronella farmers in Kedungrandu expect to provide more precise information about the features of farmers from a social and economic perspective. Socio-economic characteristics of farmers predict to illustrate what kind of farmers who cultivate citronella in Kedungrandu expect to make a positive contribution to increasing the production of citronella.

2. Materials and Methods

The research was conducted in Banyumas Regency, precisely in Kedungrandu Village, Patikraja District. The study was conducted by the survey-explanatory method. The location was determined purposively, considering that the site is a citronella-producing village, a fostered village of the Faculty of Agriculture, JenderalSoedirman University. Respondents of this study were 40 farmers who cultivated citronella in Kedungrandu Village, Patikraja District, Banyumas Regency. Respondents chose intentionally, namely farmers who did citronella farming in 2020. The method for determining respondents was using the census method, meaning that all farmers are respondents in this study. It states that the total sample takes if the population is less than 100 people [4].

3. Results and Discussion

3.1. Characteristics of Respondents by Gender

This study shows that 28 male farmers (70%) of the total respondents, 40 people. This number is higher than the number of female farmers who only numbered 12 (30%). This is because many women work to take care of the household in Kedungrandu, so that not many carry out work as farmers.

3.2. Characteristics of Respondents by Age

Table 1. Characteristics of Respondents by Age

Age	Number	Percentage
<30	1	3%
31 - 40	7	18%
41 - 60	27	68%
60>	5	13%
Total	**40**	

e
d on Table 1, farmers aged over 60 years amounted to 13%, while farmers aged between 41-60 years amounted to 68%. Based on the productive age according to BPS, which is 15-64 years, the farmers in Kedungrandu, on average, are still in productive age. However, from Table 1, it can be seen that the number of farmers aged under 40 years is much lower than other age categories. In contrast, younger farmers expect to contribute more in agriculture due to their adaptability to accept innovations.

3.3. Characteristics of Respondents Based On Education

Table 2. Characteristics of Respondents Based On Education

Education	Number	Percentage
NotCompleted Elementary School	5	13%
Elementary School	25	63%
Junior High School	5	13%
High School	4	10%
Bachelor Degree	1	3%
Total40		

Based on Table 2 farmers, 63% of citronella farmers in Kedungrandu completed their last education in elementary school. This level of education is in the low category level, where the level of formal education will affect how farmers can accept innovation, technology that will impact the productivity of their farming business. Farmers who have a higher level of education will have a better ability to make decisions to increase their productivity. The level of education of farmers does not significantly affect the level of income [5]. However, it has a positive effect on the level of income of farmers. The level of education does not affect the production of agricultural commodities but has a positive effect on the results of its output [6]. The level of education does not significantly affect income or the level of farm production. However, education has a positive influence that can show how the level of adoption of technology and information can affect farmers in running and developing their farming.

3.4. Characteristics of Respondents by Main Occupation

Citronellafarmers in Kedungranduhave a variety of primary jobs. However, as many as 35% of Citronella farmers work as farm laborers. The second highest main occupation is farmers who work as construction workers as much as 23%, followed by farmers whose main occupation is scavengers at 15%. Judging from the percentages, we can see that 65% of Citronella farmers have other primary jobs outside of agriculture. Low-income levels encourage farmers to have other jobs to meet their daily needs.

3.5. Characteristics of Respondents based on Side Jobs

Table 3. Characteristics of Respondents based on Side Jobs

Side job	Number	Percentage
Farmer	16	40%
Breeder	2	5%
None	10	25%
Brick Business	2	5%
Scavengers	3	8%
Entrepreneur	1	3%
Farmer	6	15%
Total	**40**	

Just as the primary jobs of farmers are varied, so are the side jobs of the citronella farmers in Kedungrandu. Table 3 shows that 40% of citronella farmers have a side job as farm laborers, followed by 25% of farmers who do not have a side job other than farming Citronella and 15% of citronella farmers who also work as farmers in other farming commodities. Most farmers have side jobs due to the availability of much free time from the primary entity of their farming [7]. After calculating, 75% of citronella farmers have side jobs because of economic motivation. Farmers are still looking for other side income outside of citronella farming, which expects to help meet their family's needs. Farmers have side jobs to increase their income; especially their main commodity is not happening [8].

3.6. Characteristics of Respondents based on Farming Experience

Table 4. Characteristics of Respondents based on Farming Experience

Farming Experience	Number	Percentage
<5 years	7	18%
5 - 10 years	9	23%
11 - 20 years	5	13%
>20 years	19	48%
Total	**40**	

Table 4 explains that 48% of Citronella farmers in Kedungrandu already have more than 20 years of farming experience. This figure is the highest percentage of the length of experience in farming the citronella farmers. They are then followed by farmers who have experience farming between 5-10 years and 23% and 11-20 years as much as 13%. This shows that most farmers who currently grow citronella are very experienced farmers in the agricultural world.

Farming experience divides into three categories [9]. Namely, a new venture is less than ten years, moderate experience is between 10-20 years, and 20 years old experience. This long-standing experience expects to provide many lessons and more knowledge to farmers in increasing the productivity of citronella farming. Farming experience that has been going on for a long time can also provide more learning experiences for farmers to expect to know the problems and overcome and control risks in their farming.

3.7. Characteristics of Respondents based on Experience of Cultivating Citronella

This study shows that 95% of citronella farmers in Kedungrandu have less than three years of experience cultivating citronella. This brief experience shows that the selection of citronella as a planting commodity is relatively new for most farmers; this also indicates that most citronella farmers in productive age who are old can adapt to innovations and habits, namely planting other commodities outside of their usual farming activities. Citronella also means that it is a commodity that is easy to cultivate and has a high enough value in terms of social and economic terms to provide more motivation for farmers to produce it.

Conclusions

The socio-economic characteristics of individual citronella farmers in Kedungrandu Village, Patikraja District, Banyumas Regency are male, aged 41-60 years. Most citronella farmers are educated up to elementary school (SD) and have their primary job as farm laborers. Citronella farmers mostly have side jobs as farm laborers for other commodities and, on average, have more than 20 years of experience in farming. However, the majority of citronella farmers only have less than three years of experience cultivating citronella.

References

[1] Damanik S. 2007.Analisis Ekonomi Usahatani Serai Wangi (StudiKasusKecamatanGunungHalu, Kabupaten Bandung Selatan).BuletinLiiro.18:203–21.
[2] Badan Pusat Statistik. (2021, October 10). *Indonesia Value of Essential Oil Exports. shorturl.at/bsCRU*
[3] Hapsari H, Rasmikayati E, Saefudin. 2019.KarakteristikPetani dan ProfilUsahatani Ubi Jalar di KecamatanArjasari, Kabupaten Bandung. Sosiohumaniora.21: 247 - 255
[4] Arikunto, S. 2012. ProsedurPenelitianSuatuPendekatanPraktik. RinekaCipta.
[5] Juliansyah H, Riyono A. 2018. PengaruhProduksi, Luas Lahan dan Tingkat Pendidikan terhadapPendapatanPetaniKaret di Desa Bukit HaguKecamatanLhoksukonKabupaten Aceh Utara. Jurnal Ekonomi PertanianUnimal. 1:65–72.
[6] Susanti D, Listiana NH, Widayat T. 2016. The Influence of the Farmer Ages, Levels of Education and Land Area to Blumea Yields. JurnalTumbuhanObat Indonesia. 9:75–82.
[7] Rachmadhan AA, Wibowo R, Hapsari TD. 2014.Hubungan Tingkat Kepuasan, Tingkat Motivasi dan Produktivitas Tebu Petani Mitra Kredit PG DjombangBaru. BerkalaIlmiahPertanian. 10: 1–15.
[8] Rasmikayati E, Kusumo RAB, Sukayat Y, Haikal Y, Arisyi, Saefudin BR. 2020. KarakteristikIndividu dan UsahataniPetani Mangga yang MelakukanKemitraanPemasaran di KecamatanSindangkasih, KabupatenMajalengka. ParadigmaAgribisnis. 3:12–23.
[9] Manyamsari I, Mujiburrahmad. 2014. KarakteristikPetani Dan HubungannyaDenganKompetensiPetaniLahanSempit (Kasus : Di DesaSinar Sari KecamatanDramagaKab. Bogor Jawa Barat). Agrisep.15:58–74.

Effects of rice husk biochar (RHB) with combined inoculation of arbuscular mycorrhizal fungi (AMF) and phosphate solubilizing bacteria (PSB) on growth of maize (*Zea mays*)

Aufa Ain A.S.[1] and Noraini M.J.[1*]

[1]Department of Land Management, Faculty of Agriculture, Universiti Putra Malaysia, Serdang, Selangor 43400, Malaysia

*j_noraini@upm.edu.my

Abstract. For optimal plant growth and productivity, agricultural activity is dependent on soil quality and nutrient availability. Soil amendment such as rice husk biochar (RHB), which abundantly found in Malaysia, are capable of improving soil fertility due to its longevity and other benefits to soil biota, including beneficial soil fungi and bacteria such as arbuscular mycorrhizal fungi (AMF) and phosphate solubilizing bacteria (PSB). AMF are well-known for their capacity to improve plant nutrient uptake, particularly P, which is essential for plant growth. Similarly, PSB, which is a beneficial bacteria, are able to solubilize phosphate into soluble form thus enhance the P availability for plant. However, the combination effect of AMF and PSB in the presence of RHB is not known. Thus, in order to assess the effects of AMF and PSB in the presence of RHB on soil properties and maize growth, a pot experiment was done. Randomized completely block design (RCBD) arrangement was applied to the following treatments: T1 (control), T2 (RHB), T3 (AMF), T4 (PSB), T5 (+RHB, +AMF), T6 (+AMF, +PSB), T7 (+RHB, +PSB) and T8 (+RHB, +AMF, +PSB). About 10 t/ha RHB was added to the soils for treatments T2, T5, T7, and T8. Ten grams of AMF inoculum and double application of PSB were applied to their respective treatments. Plant growth and soils sampling were done at 45 days after planting (DAP). Results showed that T7 gave 137% root biomass and 123% P uptake higher than control. Our findings were in line with our hypothesis (application of RHB, AMF or PSB or their combination would improve maize growth compared to control, in which T8 gave highest root volume and 150% higher than control. The results suggested that there were positive and synergistic effects of beneficial soil microorganisms (AMF or/and PSB) either with or without RHB. Thus, these combined inoculations and RHB application could be used in biofertilizer formulations to increase the maize growth and yield in sustainable agriculture system.

1. Introduction

Organic management of soil is vitally important for the plant growth as well as its sustainability. Biochar application, for example, is vital in sustainable agriculture, due to its slower degradation than other types of natural organic materials [19]. Biochar is normally alkaline in nature (pH>7), with the ameliorating effects on soil pH hence improving the nutrient availability. Moreover, biochar also provides habitats and increase the activity of beneficial soil microorganisms (such as arbuscular mycorrhizal fungi (AMF) and phosphate solubilizing bacteria (PSB)) which are important in nutrient cycles especially Phosphorus (P) [4, 16]. In tropical and highly weathered soils, P is the most limiting nutrient for plant uptake. The acquisition of phosphorus attributes to root development, crop maturity and production [11]. Biochar is able to interact with soil microorganisms and thus improves plant nutrient availability and increase nutrient uptake through the root [8, 20].

Symbiotic fungi such as AMF are generally known to form symbiosis relationships with more than 80% terrestrial plants and relatively non-selective [15]. Mycorrhizal associations through its several roles as biofertilizer were beneficial for soil and plant growth. The most outstanding roles of AMF is on P acquisition that the extraradical mycelia move farther to obtain the nutrient. Plant inoculated with AMF also acquire water and nutrients more efficient. AMF also had a positive impact on soil properties (physical and chemical) through a network of hyphae within the soil and between plant [17, 20]. Besides AMF, a group of beneficial bacteria known as PSB can also help AMF by forming synergistic relationship with plants. The bacteria provide soluble phosphate while the plant supply carbon compound. Generally, PSB plays an important function in P nutrition by exchanging its availability to plants by solubilizing and mineralizing phosphorus from organic soil P pools into inorganic form [17]. The mechanisms involved through microbial generation of organic acids and mineralization of organic phosphorus by acid phosphates [3].

While it is well established that PSB and AMF could improve P availability in soil, their combinations in the presence of biochar is limitedly reported. Studies on rice husk biochar (RHB), AMF and PSB however raised few concerns and aspects to be considered. The concerns include; i) the abundance of AMF affected by the biochar application and PSB inoculation. In P deficient soil, the hyphae structure (to boost the plant P uptake) may be reduced due to P concentration increases as PSB improve the soil properties that alter the P concentration in the soil solution. There were several reports on biochar affect the AMF abundance while some report otherwise [9,10]. Moreover, ii) the PSB efficiency to enhance soluble P for maize nutrient uptake when AMF and RHB are present. Bacteria inoculation increased AMF root colonization. Under insufficient nutrient supply, high AMF colonization inhibit P solubilizers. As a result, the bacteria become less significant and the plant dependent more to AMF [20, 22]. In line with the concerns above, the combined effects of biochar which are abundance and locally available such as rice husk biochar (RHB), AMF and PSB on maize growth is not known especially when applied to Malaysian highly weathered soils. The information on these interactions is limited. Thus, this experiment was conducted with the objective to determine the effects of AMF and PSB in the presence of RHB on maize growth.

2. Materials and methods

2.1. Experimental design, planting preparation and treatment application

The pot experiment was carried out for 45 days at rain shelter Ladang 15, Faculty of Agriculture, Universiti Putra Malaysia. The treatments involved were 0, 10 t/ha rate of RHB, with/ without AMF, and with/without PSB in randomized complete block design (RCBD) arrangement with four replicates. Overall, there were eight treatments; T1 (control), T2 (RHB), T3 (AMF), T4 (PSB), T5 (+RHB, +AMF), T6 (+AMF, +PSB), T7 (+RHB, +PSB) and T8 (+RHB, +AMF, +PSB). In general, there were 32 experimental units.

Rice husk biochar (RHB) was collected from BERNAS (Malaysian National Rice Agency) rice mill in Tanjung Karang, Selangor. It consists 23.26% C, 0.38% N and 0.05% P with pH 6.94. About 5 kg of soils (Serdang series having 1.40% C, 0.14% N and -0.05% P) were mixed thoroughly with RHB for designated treatments and laid in polybags (12" x 12").

Prior to the experiment, the AMF inoculum was prepared. The AMF spores (*Glomus mosseae*) obtained from the Department of Land Management's Soil Microbiology Laboratory were cultured in sterile sand with a maize plant as a host. Treatment with AMF inoculation, about 10 g of AMF inoculum

(>50 AMF spores) were placed into a hole in the polybag's centre underneath the maize seeds (3 seeds per polybag). While non-AMF treatments received equal amount (10 g) of sterile sand.

The PSB inoculum was prepared using bacterial PSB16 (*Bacillus spp.* strain). In each bacterial treatment, an inoculum of approximately 5 x 108 cfu ml^{-1} live washed bacterial cells was employed [13]. The amount given to the non-inoculated pots was the same as the inoculated pots, but without the inoculum. The optical density of the inoculum was measured prior to application for optimum bacterial growth. The bacterial cells were applied two times, first during seed sowing and the second times two weeks after sowing

2.2. Determination of plant growth measurement and soil nutrient analysis

The plants were harvested on 45 days after planting. The plant upper (shoot) and bottom parts (roots) were separated for weighing and analysis. Plant samples were washed to eliminate all soil particles and dried in a 70°C oven for 3 days until they reached a consistent weight. Nutrient analysis was done by digesting the plant tissue and air-dried soil based on Dry Ashing Method (Kalra, 1997). The available phosphorus was determined by using Bray 2 method (Bray and Kurtz, 1945).

2.3. Statistical analysis

All collected data were statistically analysed using Statistical Analysis System (SAS) version 9.4. Tukey's HSD (Honest Significant Difference) at $P \leq 0.05$ was used to perform the means separation.

3. Results and discussion

3.1. Plant growth measurement, soil P and plant P uptake

Generally, there were significant effects of the treatments on the plant growth parameters. Treatments with RHB (10 t/ha) had significantly ($P \leq 0.05$) higher plant height, shoot and root biomass as well as root volume. Based on Table 1, treatment with RHB alone, T2 (10 t/ha RHB), which produced the highest plant height was significantly ($P \leq 0.05$) higher than T1 (control), T3, (AMF alone) and T7 (AMF + PSB). However, T2 but did not differ significantly ($P \leq 0.05$) to other treatments added with RHB (that of T4, T5, T6 and T8). This showed that the effect could be highly attributed to biochar itself rather than the beneficial microbes (T3 and T7). Meanwhile, treatment inoculated with beneficial microbes (T5, T7 and T8) significantly influenced and resulted in highest shoot and root dry weight as well as root volume, respectively. This indicated that each beneficial microbes, either AMF or combinations (AMF, +RHB) performed their roles to provide a conducive environment for best plant growth. Control soil had the lowest values and poor plant growth among all treatments.

Table 1. Plant growth measurement 45 days after planting

Treatment	Plant height (cm)	Plant biomass (Dry weight) (g)		Root volume (cm^3)
		Shoot	Root	
T1 (Control)	55.00b	1.21b	0.700b	3.33c
T2 (RHB)	80.00a	2.82ab	2.86ab	12.00bc
T3 (AMF)	54.25b	2.45ab	0.97b	13.00abc
T4 (PSB)	63.50ab	1.60ab	2.71ab	12.00bc
T5 (RHB + AMF)	63.75ab	3.76a	2.15ab	18.00ab
T6 (RHB + PSB)	66.33ab	1.27b	1.43ab	14.00abc
T7 (AMF + PSB)	48.00b	2.21ab	3.79a	16.00ab
T8 (RHB + AMF + PSB)	71.25ab	2.94ab	2.05ab	22.75a
Pr > F	0.0062*	0.0145*	0.0132*	0.0013*

Note: means with the same letter in a column are not significantly different between treatments at P≥0.05 according to Tukey's HSD (Honest Significant Difference). Pr>F with * show significant at P≤0.05 (ANOVA).

There was significant effect (P≤0.05) of soil P concentrations as shown in Table 2, which only exhibited on treatment with AMF inoculation (T3). The significantly (P≤0.05) highest soil P in T3 compared to other treatments indicated the outstanding role of AMF in solubilizing P, which was not achieved by PSB alone (T4). Significant effects (P≤0.05) were shown on total P uptake by plant (Table 2). Treatments with combined inoculation of AMF and PSB (T7) had significantly shown the best total P uptake and significantly (P≤0.05) comparable to control. The presence of both AMF and PSB although was not obvious in the P solubilization (plant P), their combination somehow able to influence the P uptake by plants.

Table 2. Soil P and plant P uptake 45 days after planting

Treatment	Soil Phosphorus (mg kg^{-1})	Plant P uptake (mg $plant^{-1}$)
T1 (Control)	0.58b	1.23b
T2 (RHB)	1.38b	4.05ab
T3 (AMF)	3.1a	2.60ab
T4 (PSB)	2.43b	2.50ab
T5 (RHB + AMF)	2.45b	3.67ab
T6 (RHB + PSB)	2.01b	1.50ab
T7 (AMF + PSB)	1.72b	5.13a
T8 (RHB+AMF + PSB)	1.05b	4.00ab
Pr > F	<.0001*	0.0378*

Note: means with the same letter in a column are not significantly different between treatments at P≥0.05 according to Tukey's HSD (Honest Significant Difference). Pr>F with * show significant at P≤0.05 (ANOVA).

3.2. Discussion

The inoculation of AMF had significant effects on maize growth and soil properties in terms of plant height, plant total biomass and root volume. One of the important structures of AMF, which is the hyphae, would benefit the plant in altering the soil pH and nutrient acquisition that directly improved the plant growth [5, 20]. In addition, AMF has a prominent role to solubilize P in soil, where the element is believed to be the second limiting factor in maize growth [2]. Other than that, AMF influenced the P uptake by the plant as shown by the result of combined inoculation of AMF and PSB. It showed that the maize plants were utilizing the inorganic form of P. For microbial aspects, the treatment of RHB and PSB affected the sporulation and abundance of AMF. From the results, number of total spore count and root infection gave significant differences according to the treatments. RHB through their porous characteristics would provide shelter for AMF; hence increase in microorganisms' survival-population and activities in soils [7] (data not shown). The combined inoculation of AMF and PSB resulted in better soil P concentration and P uptake by plant. PSB would have positive effect on AMF development, which acts as mycorrhizal helper bacteria to initiate the growth of hyphae and better colonization. In addition, PSB also would benefit the performances of AMF, which the PSB solubilize the fixed P, then transferred to the plant root system via the external AMF hyphae, boosting mineral P acquisition [21]. From the results, the function of AMF was more significant to improve soil properties especially P concentration in soil.

PSB is a group of beneficial bacteria that would improve the sparingly soluble P to be available for plant [14]. From the results, the treatment with PSB gave insignificant effects on plant growth, however showed better results in improving soil P concentration and plant uptake. The performances of PSB were significant with combination of AMF and RHB. There were shown by the of soil P concentration, which treatment of PSB alone gave lower values compared to treatment with AMF alone. The less significant role of PSB would possibly due to initial soil P status and outstanding performances of AMF in P acquisition. Under insufficient nutrient supply, high AMF colonization inhibits P solubilizes. As a result, the bacteria become less significant and the plant dependent more to AMF [22]. Moreover, the PSB function as helper to initiate the development of AMF hyphae in line with previous findings, there were very strong effects of bacteria strain on the growth of AMF extra radical hyphae outside the roots of carrot and on colonization of potato roots by AMF [12]. Treatment with PSB and the addition of biochar, the soil environment and soil microorganisms are interacted each other, thus promoting the improvement of soil microbial ecological system together [23].

RHB was applied with 10 t/ha, recommended to improve plant growth [1]. This is similar to the findings that biochar improved height of oat plant [18] and growth performance of soursop seedlings [6]. Biochar through its indirect mechanisms such as soil amelioration could provide a conducive environment for plant growth. Their presence in soils affects the structure, texture, porosity, particle size distribution, and density of the soil. Biochar's molecular structure is chemically and microbially stable. Most biochar has a porous structure and a large surface area, which is a significant physical property. This structure affects the binding of essential nutritive cations and anions, allowing for greater water retention and nutrient retention. [1] and increase microbial activity [8]. In addition, RHB with AMF treatment showed higher number of total spore count and root infection, which RHB provide better soil environment for AMF development by altering the soil pH and providing shelter [20].

4. Conclusions

In conclusion, the study showed that arbuscular mycorrhizal fungi (AMF) and phosphate-solubilizing bacteria (PSB) as single inoculation (T3, T5) or combined inoculations with or without rice husk biochar (RHB) as found in T7 and T8 improved maize growth and P nutrition (in soils and plant uptake). Their treatments showed significant higher in plant height, root and shoot biomass, root volume, soil P and plant P uptake compared to control. The treatment with RHB and PSB had significantly increased the abundance of AMF, which represent by higher number of total spore count and percentage of positive root infection. From the experiment, it can be concluded that soil treated with RHB and inoculated with both AMF and PSB (T8) was the most effective in improving root volume (for more water and nutrient acquisitions) as appeared to be the underlying mechanism or probable reasons in better P concentration in soils hence plant P uptake compared to control.

References

[1] Abdulrazzaq, H., Jol, H., Husni, A., & Abu-Bakr, R. (2014). Biochar from empty fruit bunches, wood, and rice husks: effects on soil physical properties and growth of sweet corn on acidic soil. *Journal of Agricultural Science*, *7*(1), 192.

[2] Almethyeb, M., Ruppel, S., Paulsen, H. M., Vassilev, N., & Eichler-Löbermann, B. (2013). Single and combined applications of arbuscular mycorrhizal fungi and Enterobacter radicincitans affect nutrient uptake of faba bean and soil biological characteristics. *Agric. Fores. Res*, *3*, 229-234.

[3] Anand, K. U. M. A. R., Kumari, B. A. B. Y., & Mallick, M. A. (2016). Phosphate solubilizing microbes: an effective and alternative approach as biofertilizers. *J Pharm Pharm Sci*, *8*(2), 37.

[4] Bruun, S., Clauson-Kaas, S., Bobuľská, L., & Thomsen, I. K. (2014). Carbon dioxide emissions from biochar in soil: role of clay, microorganisms and carbonates. *European Journal of Soil Science*, *65*(1), 52-59.

[5] Gosling, P., Hodge, A., Goodlass, G., & Bending, G. D. (2006). Arbuscular mycorrhizal fungi and organic farming. *Agriculture, Ecosystems & Environment*, *113*(1), 17-35.

[6] Harun, N. S. N., Jaafar, N. M., & Sakimin, S. Z. (2021). The effects of rice husk biochar rate on arbuscular mycorrhizal fungi and growth of soursop (Annona muricata L.) seedlings. Sustainability, 13(4), 1817.

[7] Jaafar, N. M. (2014). Biochar as a habitat for arbuscular mycorrhizal fungi. In Mycorrhizal Fungi: Use in Sustainable Agriculture and Land Restoration (pp. 297-311). Springer, Berlin, Heidelberg.

[8] Jaafar, N. M., Clode, P. L., & Abbott, L. K. (2015). Soil microbial responses to biochars varying in particle size, surface and pore properties. *Pedosphere*, *25*(5), 770-780.

[9] Liu, S., Tang, W., Yang, F., Meng, J., Chen, W., & Li, X. (2017). Influence of biochar application on potassium-solubilizing Bacillus mucilaginosus as potential biofertilizer. *Preparative Biochemistry and Biotechnology*, *47*(1), 32-37.

[10] Matsubara Y-I, Hasegawa N, Fukui H (2002) Incidence of Fusarium root rot in asparagus seedlings infected with arbuscular mycorrhizal fungus as affected by several soil amendments. J Jpn Soc Hortic Sci 71:370–374

[11] Mohammadi, K. (2012). Phosphorus solubilizing bacteria: occurrence, mechanisms and their role in crop production. *Resour Environ*, *2*(1), 80-85.

[12] Ordoñez, Y. M., Fernandez, B. R., Lara, L. S., Rodriguez, A., Uribe-Vélez, D., & Sanders, I. R. (2016). Bacteria with Phosphate Solubilizing Capacity Alter Mycorrhizal Fungal Growth Both Inside and Outside the Root and in the Presence of Native Microbial Communities. *PLoS ONE*, *11*(6), e0154438.

[13] Panhwar, Q. A., Othman, R., Rahman, Z. A., Meon, S., & Ismail, M. R. (2012). Isolation and characterization of phosphate-solubilizing bacteria from aerobic rice. *African Journal of Biotechnology*, *11*(11), 2711-2719.

[14] Panhwar, Q. A., Radziah, O., Zaharah, A. R., Sariah, M., & Razi, I. M. (2011). Role of phosphate solubilizing bacteria on rock phosphate solubility and growth of aerobic rice. *Journal of Environmental Biology*, *32*(5), 607.

[15] Piotrowski, J. S., Morford, S. L., & Rillig, M. C. (2008). Inhibition of colonization by a native arbuscular mycorrhizal fungal community via Populus trichocarpa litter, litter extract, and soluble phenolic compounds. *Soil Biology and Biochemistry*, *40*(3), 709-717.

[16] Rafique, M., Sultan, T., Ortas, I., & Chaudhary, H. J. (2017). Enhancement of maize plant growth with inoculation of phosphate-solubilizing bacteria and biochar amendment in soil. Soil science and plant nutrition, 63(5), 460-469.

[17] Saranya, K., & Kumutha, K. (2011). A critical review on arbuscular mycorrhizal fungi and soil properties in natural systems. *Int. J. Curr. Res*, *33*(5), 032-038.

[18] Schulz, H., Dunst, G., & Glaser, B. (2013). Positive effects of composted biochar on plant growth and soil fertility. Agronomy for sustainable development, 33(4), 817-827.

[19] Spokas, K. A., Baker, J. M., & Reicosky, D. C. (2010). Ethylene: potential key for biochar amendment impacts. *Plant and Soil*, *333*(1-2), 443-452.

[20] Warnock, D. D., Lehmann, J., Kuyper, T. W., & Rillig, M. C. (2007). Mycorrhizal responses to biochar in soil–concepts and mechanisms. *Plant and soil*, *300*(1-2), 9-20.

[21] Widada, Jaka & I. Damarjaya, D & Kabirun, S. (2007). The interactive effects of arbuscular mycorrhizal fungi and rhizobacteria on the growth and nutrients uptake of sorghum in acid soil. Development in Plant and Soil Science. 102.

[22] Wu, S. C., Cao, Z. H., Li, Z. G., Cheung, K. C., & Wong, M. H. (2005). Effects of biofertilizer containing N-fixer, P and K solubilizers and AM fungi on maize growth: a greenhouse trial. Geoderma, 125(1), 155-166.

[23] Yan-li, D. I. N. G., Jie, L., & Ying-ying, W. A. N. G. (2013). Effects of biochar on microbial ecology in agriculture soil: A review. *Yingyong Shengtai Xuebao*, *24*(11).

IOP Conf. Series: Earth and Environmental Science 1131 (2023) 012008 doi:10.1088/1755-1315/1131/1/012008

Healthy organic coconut sugar powder business's development strategy: A case study at the Nira Perwira Cooperative, Purbalingga District, Central Java, Indonesia

Budi Dharmawan*, Suyono, Agus Sutanto, Irene Kartika Wijayanti, Dindy Darmawati Putri

Universitas Jenderal Soedirman, Purwokerto, Indonesia

*budi.dharmawan@unsoed.ac.id

Abstract. Efforts to establish the Nira Perwira cooperative to increase the added value of coconut sugar powder through better processing and export orientation are still hampered by a number of challenges. Uncertainty in supply and the quality of supply from partner farmers are two of the difficulties encountered. These research objectives are to 1) analyze the external and internal factors of the Nira Perwira Cooperative in developing the organic coconut sugar powder business; 2) analyze the Nira Perwira Cooperative's position in the organic coconut sugar powder; and 3) analyze alternative strategies that can be used in developing the Nira Perwira Cooperative's organic coconut sugar powder. The study makes use of both secondary and primary data gathered through interviews, observation, recording, and Focus Group Discussions (FGD). The case study method was used in this study. Internal Factor Evaluation (IFE), External Factor Evaluation (EFE), Internal-External (IE), Strength, Weakness, Opportunities, and Threats (SWOT) were used to analyze the collected data. The results revealed that the IFE value was 2.28 and the EFE value was 3.06, indicating that the cooperative position is in quadrant II of the IE matrix. Based on the SWOT analysis, the Nira Perwira Cooperative's organic coconut sugar powder development strategy should include more promotional activities, more farmer loyalty, more contracts with third parties, more marketing networks, more competent members to run marketing, and more internal reorganization.

Keywords: *Development strategy, organic coconut sugar powder, Nira Perwira Cooperative.*

1. Introduction

The low price of agricultural products at the producer level cannot be separated from the less competitive structure of the market for farm products. On the one hand, some parties are more dominant in determining the price (price makers). On the other hand, producers' bargaining positions are weak. Producers tend to be price takers. Zdiniakova and de la Calle [1] reported that coconut sugar marketing tends to be less efficient. The coconut sugar market structure at the whole seller level is loose oligopsony, while at the collector and farmer level, it is oligopoly. Hence, the buyer determines the price of coconut sugar at the farm level.

Purbalingga District is one of Central Java's coconut sugar production centers. In 2015 coconut sugar production in this district reached 53,130,80 tons [2], involving around 18,000 units of coconut sugar farmers. The economy of most farmers or *penderes* is still a concern [3].

Related to efforts to improve the welfare of farmers, Non-Governmental Organizations (NGOs) Research and Development Institute for Resources and the Environment (LPPSLH) Purwokerto have assisted coconut sugar farmers in producing organic coconut sugar powder, which has a higher selling value. At the time of pioneering assistance, the price of printed coconut sugar at the farmers' level was Rp7,000 to Rp7,500 per kg, while the cost of organic coconut sugar powder was Rp10,000 to Rp11,000 per kg, with a price difference between Rp2,500 and Rp4,000 per kg, which is very helpful for increasing the income of farmers.

In 2022, the Nira Perwira Cooperative will be the shelter for four villages in Purbalingga that produce organic coconut sugar powder. A total of 401 organic coconut sugar powder farmers were certified by the Dutch Control Union Institute in May 2014. The programs offered by the Nira Perwira Cooperative to organic coconut sugar powder farmers include (1) savings; 2) remaining operating results; (3) production support facilities; (4) market access; (5) production quality assurance; (6) information supporting the development of organic palm sugar; (7) coconut farmers' compensation fund in the form of community insurance; and (8) currently working on health insurance or BPJS (accidents, work, retirement). As a center for coconut sugar production in Purbalingga District, many farmer's families in Kutasari District reach no less than 7,000 families. Most coconut sugar producers do not sell their products to final consumers but local village collectors. Furthermore, the collectors sell them back to wholesalers in the local village or sub-district area. By wholesalers, printed coconut sugar products are sold outside the city either to meet soy sauce factories in big cities or to meet the needs of end consumers through traditional retailers in small markets in big cities. In general, producers are also bound by a down payment from buyer traders, so they will sell the products they produce to traders who have given loan money at a dominant price determined by the buyer.

Several internal and external factors influence the Nira Perwira Cooperative's critical role in improving the welfare of organic coconut sugar powder farmers. The cooperative's internal and external factors must be adequately identified in order to develop a strategy that will support the successful development of the Nira Perwira Cooperative's organic coconut sugar powder business. The success of the organic coconut sugar powder business is due to implementing the right strategy that will benefit the cooperative and help the organic coconut sugar powder farmers. In other words, the cooperative's goal, namely the welfare of organic coconut sugar powder farmers, will be realized.

2. Material and Methods

2.1. Material

The research was carried out at the Nira Perwira Cooperative, a cooperative with specific superior products of organic coconut sugar powder, and its products have met international organic standards. The research technique used is a case study with a qualitative approach. Determination of critical informants for determining strategic factors, determining weight and attractiveness is done purposively. Critical informants for strategic factors are the Head of the Nira Perwira Cooperative, Head of Quality Assurance Division, Manager, Head of Production Division, Head of Logistics, and Supervisory Division. The critical informant to determine the attractiveness value is the Chairman of the Nira Perwira Cooperative.

2.2. Methods

This study necessitates the collection of both primary and secondary data. Preliminary data includes the state of the Nira Perwira Cooperative, an overview of the cooperative's organization, and information on the cooperative's marketing activities of organic coconut sugar powder to exporters. Secondary data is already available at research-related organizations such as the Nira Perwira Cooperative, the Purbalingga Regency Central Bureau of Statistics, and the Department of Trade. Interviews, observation, recording, and focus group discussions are all methods for gathering data (FGD). IFE (Internal Factor Evaluation) and EFE (External Factor Evaluation) analyses of internal and external factors using matrix analysis (External Factor Evaluation). The Nira Perwira Cooperative's position in the organic coconut sugar powder industry is determined using the IE (Internal External) matrix.

3. Results and discussion

3.1. Internal and external factors of the Nira Perwira Cooperative

Internal factors were identified, resulting in the inner strengths and weaknesses of the Nira Perwira Cooperative. Table 2 shows the Nira Perwira Cooperative's strength and weakness factors. The identification of external factors demonstrates the Nira Perwira Cooperative's external opportunities and threats; the identification of external factors is shown in Table 1.

Table 1. Identification of the internal and external factors of the Nira Perwira Cooperative

No.	Internal Factors	Strengths	Weaknesses
1	**Production**	a. Using ingredients that are safe for health b. Continuous supply of coconut sugar c. Prices at the farm level are stable	Inability to produce large quantities on demand
2	**Marketing**		Sales on a large scale are still in scope around the district, and exports are not sustainable
3	**Financing**		Weak capital
4	**Management**	a. Have a legal entity b. The existence of training to improve the quality of human resources c. Have a quality assurance body	a. Less active member participation b. Management is lacking in excellent knowledge
	External Factors	**Opportunities**	**Threats**
1	**Macro Environment**		
	Economic factors		Fluctuating consumer demand
	Technological Factor	Online marketing allows purchases made anytime and from anywhere	
	Politics, government, law	Support from district government through development programs	
	Socio-cultural	a. The number of coconut sugar farmers is large b. Trend to organic consumption c. Have good relationship among farmer, trader, government	Lack of public awareness and understanding of cooperatives
	Climate condition		Unpredictable rainy season
2	**Industrial Environment**		
	Competition in the industry		Many new organic coconut sugar powder industries have sprung up
	Potential buyers	Many new requests from potential buyers abroad	

Source: Primary Data Analysis (2021).

3.2. Analysis of the position of the Nira Perwira Cooperative in the organic coconut sugar powder industry

The IFE value of the Nira Perwira Cooperative is 2.28, and the EFE value is 3.06. An IFE value below 2.50 indicates that the Nira Perwira Cooperative is internally weak. The EFE value above 2.50 means that the Nira Perwira Cooperative can explore opportunities and avoid threats [4].

Table 2. Internal Factor Evaluation (IFE) and External Factor Evaluation (EFE) Matrices

Main Internal Factors	Weight	Rating	Score
Strengths			
1. Have a legal entity	0.07	3	0.21
2. Using ingredients that are safe for health	0.10	4	0.40
3. Continuous supply of coconut sugar	0.09	4	0.36
4. Prices at farm level are stable	0.08	3	0.24
5. Have a quality assurance body	0.09	3	0.27
6. Training to improve the quality of human resources	0.07	3	0.21
Weaknesses			
1. Less active member participation	0.08	1	0.08
2. Unable to produce in large quantities as order	0.09	2	0.18
3. Sales on a large scale are still in scope around the district, and exports are not sustainable	0.09	1	0.09
4. Weak capital	0.09	1	0.09
5. Less active member participation	0.07	1	0.07
6. Management is lacking in excellent knowledge	0.08	1	0.08
Total	1.00		2.28
Main External Factors	**Weight**	**Rating**	**Score**
Opportunities			
1. The number of coconut sugar farmers is large	0.09	2	0.18
2. Cooperation network with facilitators/ partners is well established	0.11	4	0.44
3. Many new requests from potential buyers abroad	0.10	4	0.40
4. There is support from the district government through development programs	0.09	3	0.27
5. Online marketing allows purchases to be made anytime and from anywhere	0.09	3	0.27
6. People's trend towards organic consumption	0.12	4	0.48
Threats			
1. Fluctuating consumer demand	0.09	2	0.18
2. Unpredictable rainy season	0.09	2	0.18
3. Lack of public awareness and understanding of cooperatives	0.10	3	0.30
4. Many new organic coconut sugar powder industries have sprung up	0.12	3	0.36
Total	1.00		3.06

Source: Primary Data Analysis (2021).

The IFE and EFE matrices' values are used as inputs in the IE matrix, so it is known that the position of the Nira Perwira Cooperative in the organic coconut sugar powder industry is in position II. Position II is a position of growth, development, and cooperatives. The strategies that can be applied are integrative and intensive.

		IFE *Good* 3.00-4.00	*Average* 2.00-2.99	*Weak* 1.00-1.99
EFE	*Good* 3.00-4.00	I *Growth and Build*	II *Growth and Build*	III *Hold and Maintain*
	Average 2.00-2.99	IV *Growth and Build*	V *Hold and Maintain*	VI *Harvest and Divest*
	Weak 1.00-1.99	VII *Hold and Maintain*	VIII *Harvest and Divest*	IX *Harvest and Divest*

Figure 1. Internal-External (IE) Matrix

3.3. The alternative strategy of organic coconut sugar powder's business development

Alternative strategies that follow the position of the Nira Perwira Cooperative as follows:

a. SO strategy

Increase promotional activities

The Nira Perwira Cooperative should use the internet as a promotional means. The strategy that can be done is to create a website that contains Nira Perwira Cooperative's complete profile, achievements, products produced, question and answer column, and procedures for ordering products.

b. ST strategy

b.1. Increase farmer loyalty

The Nira Perwira Cooperative must show actual performance so that organic coconut sugar powder farmers have confidence in working with the Nira Perwira Cooperative. The steps that can be taken are to consistently fulfil the rights and needs of coconut sugar farmers.

b.2. Make a contract with a third party

The strategy to make a profitable contract for the Nira Perwira Cooperative includes the contract period, number of products, and price.

b.3. Improve product competitiveness

The quality of existing products must be maintained and improved through supervision from the quality assurance sector. As a result, customers are eager to repurchase Nira Perwira Cooperative products.Product diversification can also increase product competitiveness [5].

c. WO strategy

c.1. Improve marketing network

The strategy to promote marketing partners is to establish relationships with other parties to support each other [6]. Cooperatives must be selective and observant in choosing partners, and this will ensure that the partnerships that are built are mutually beneficial [7]. The Cooperative Service, which is part of the district government, has a role in recommending the Nira Perwira Cooperative to other people.

c.2. Recruiting competent members to manage the marketing field

The lack of human resources to manage marketing activities impacts marketing activities [8]. In light of the growing awareness and entrepreneurial spirit of cooperative managers, cooperatives should have competent resources in marketing.

d. WT strategy
d.1. Carrying out Internal Reorganization
Internal reorganization is to make irresponsible and incompetent individuals replace them with better individuals [9]. In addition, new employees in the organizational structure can now spend more time on their jobs because of the internal restructuring.
d.2. The priority of organic coconut sugar powder's business development strategy
Several alternative strategies were selected from the SWOT matrix analysis, which was considered in the IE matrix analysis [10]. The appropriate alternative strategy is to increase promotional activities, increase product competitiveness, and recruit competent members to manage marketing.

4. Conclusion

Uncertainty in supply and quality of supply from partner farmers are some of the challenges faced. The results revealed that the IFE value was 2.28 and the EFE value was 3.06, indicating that the cooperative position is in quadrant II of the IE matrix (growing and fostering). Based on a SWOT analysis, alternatives for developing the palm sugar business strategy of the Nira Perwira Cooperative based on a SWOT analysis include increasing promotional activities, increasing the loyalty of coconut sugar farmers, entering into contracts with buyers, increasing marketing networks, recruiting competent members to manage marketing, increasing product competitiveness, and internal reorganization. The best thing to do in the Nira Perwira Cooperative is to make their products more competitive. The Nira Perwira Cooperative developing the organic palm sugar business must coordinate to improve internal weaknesses and be responsive to all current opportunities and threats. The Nira Perwira Cooperative must improve competitiveness through action and quality improvement to achieve a competitive advantage. The cooperative must add a marketing division to its organizational structure.

References

[1] Zdiniakova, T., and M. B. de la Calle. 2020. Feasibility study about the use of element profiles determined by ED-XRF as screening method to authenticate coconut sugar commercially available. European food research and technology. 246(10): 2101-2109.
[2] BPS Banyumas District. 2021. Banyumas District in Number. BPS Banyumas District, Central Java.
[3] Purnomo, H. 1992. Sugar components of coconut sugar in Indonesia. ASEAN Food Journal. 7(4): 200-201.
[4] David, F. R. 2003. Strategic management case writing: Suggestions after 20 years of experience. SAM Advanced Management Journal. 68(3): 36.
[5] Persaud, N., M. Woodhouse-Jackson, and M. Scriven. 2016. Enhancing the strategic management process through the use of professional evaluation methods and the logic of evaluation. Journal of multidisciplinary evaluation. 12(26): 1-11.
[6] Newbert, S. L., R. J. David, and S. K. Han. 2014. Rarely pure and never simple: Assessing cumulative evidence in strategic management. Strategic Organization. 12(2): 142-154.
[7] Bakar, A. H. A., M. A. Tufail, M. N. Y. M. A. Tufail, and W. Virgiyanti. 2011. Implementation of strategic management practices in the Malaysian construction industry. Pakistan Journal of Commerce and Social Sciences (PJCSS). 5(1): 140-154.
[8] Jennings, D. 2002. Strategic management: an evaluation of the use of three learning methods. Journal of Management Development. 27: 20-25.
[9] Jennings, D. 2002. Strategic management: an evaluation of the use of three learning methods. Journal of Management Development. 27: 20-25.
[10] David, M. E. and F. R. David. 2017. The quantitative strategic planning matrix: a new marketing tool. Journal of strategic marketing. 25(4): 342-352.

Development Model for Sustainable Utilization of Among Tani Application (A Study in Batu Smart City)

Alia Fibrianingtyas*, Vi'in Ayu Pertiwi, Neza Fadia Rayesa

Universitas Brawijaya, Malang, Indonesia

*alyaftyas@ub.ac.id

Abstract. The extention in the industrial revolution 4.0 greatly changes and affect in disemination of information methods that should be delivered to farmers toward more dynamic [1]. Information and communication technology in agricultural development requires a certain educational process and capacity building due to some technical difficulties and lack of skills to implement it [4]. In Batu City, the older farmers can not get the beneficial from smart city concepts. They difficult to use the Among Tani application. So, this research important to design the development model scenario for sustainable use of Among Tani Application in Batu City. This research use qualitative method. The data were collected from indepth interview and complemented with secondary data. These data analyzed by descriptive method. The results of this research describe about the stakeholder role in accompaniment utilization of Among Tani and it obstacles. Beside that, the development model to made it sustain were created. So, it can be conclude that the job description of farmer facilitator ought to added, especially to help the digital marketing process for farmers.

1. Introduction

Agriculture was a sector that did not escape the impact of the industrial era 4.0, especially in extension. Changed in the strategic environment greatly affect the way in which information and dissemination methods were delivered to farmers with a wider target range. These changed limit the task and function of extension, which from the beginning functioned as a carrier of technological information for innovation to occur, has changed towards providing a more dynamic flow of information [1]. Farmer's information sources could be easily accessed anywhere, started from aspects of technology, institutions, availability of logistics for production and marketing.

The openness of access to information was the result of the effects of the use of information and communication technology, particularly the advancement of mobile phones and the internet. The emergence of new variations of communication media that are integrated with each other, resulting in reduced boundaries between media [2, 3]. Information and communication technology in agricultural development required a certain educational process and capacity building due to some technical difficulties and lack of skills to implement it [4].

Agriculture in Indonesia is dominated by small farmers who have several limitations, such as low capital, land tenure, skills, knowledge, accessibility of market information and agricultural technology. The low modernity or the ability of farmers to open themselves to a renewal will worsen the condition of farmers in rejected or accepted innovations. This will lead to low incomes and conditions that are difficult to develop [5].

Some of these factors have resulted in the agricultural industrial revolution 4.0 so far have not been successful. For example, in 2017 the Batu City government provided the Batu Among Tani Teknologi (BATT) platform to improve agricultural services by utilizing android mobile phones. However, in Batu City the majority of the farmers's age were over 50 years old, so they have difficulty to accepted this innovation [6].

There were gaps and broken relationships (missing links) between the 'millennial' generation and the 'predecessor' generation. From the results of the Susenas between BPS and Kemenpppa in 2017, it is known that the millennial generation is 33.75% or around 88 million people [7]. There are values and social capital that are not connected. In his research [8], stated that social capital plays an important role in bridging people with broken relationships. To faced and addressed the gaps in social relations that were getting sharper due to the application of technology, it was necessary to play the role of community development actors (extensions, facilitators, or empowerers) to create a socio-ecosystem balance [9].

Community development actors must interferred with groups that were discriminated against and marginalized by technology. The role of farmer facilitator in this case was to bridge, bond, and built networks (linking). The three social capitals were the main spirit of social values for sharing, collaborating and participating. Community development actors must be creative and innovative in created a balance of socio-ecosystems. Hopefully that all levels of society can improve welfare and have competitiveness in the era of the industrial revolution 4.0.

2. Materials and Methods

2.1. Material. This study uses a qualitative approach. The study began with a preliminary survey and literature review to identify problems and analyze the existing farmers's demographic conditions in Bumiaji Sub-District. At this stage, it produced a real conditions of the problem and the object of research consisting of several actors (stakeholders) who involved in the modeling. In addition, each actor (stakeholder) identified their role and contribution in counseling and empowerment activities. These data were obtained through indepth interviews with key informants, namely Mr. Andik, as the Coordinator of Farmer Facilitator. The next informants were came from several Farmer Facilitators, Department of Agriculture Worker, Extention Workers, and farmers.

The research location was determined purposively in Batu City. Batu City was chosen because it currently applying Batu Smart City Master Plan. However, there was a gap between the millennial generation and the previous generation, so that the utilization of Among Tani application in Batu City have not been successful. So, researchers thought that Bumiaji Sub-District was the best location to choosen because facilitators in Bumiaji sub-district gave the fastest responses to describe any innovations that have been applied to follow up Among Tani application in marketing strategy.

2.2. Methods

The sampling method that used in this study was snowball sampling. Snowball sampling is a technique for determining samples that are initially small in number, then enlarge [10]. Firstly, researchers met key informants namely Mr. Andik as a Coordinator of Farmer Facilitator in Bumiaji Sub-District. He recommended researcher to met Extension Workers, Department of Agricultural Workers, several Farmer Facilitators, and also Farmers. So that the number of informants obtained were 11 people, the majority of whom were farmers.

The data were collected using in-depth interviews with each community development actors in Bumiaji Sub-District to formulate an effective extension modeling scenario to achieve socio-ecosystem balance. In-depth interviews were also conducted with the farmer facilitator to confirm the results of the community development actors. We used an interview guideline during the interviews. In addition, secondary data was also needed to be used as a reference in wrote research results from various sources, such as village potential books, village websites, and several relevant articles.

The data that have collected been analyzed by used interactive model's Miles and Hubberman theory. There were three steps in this theory, they are data reduction, data display, and data conclusion drawing/verification [10].

1. Data Reduction

Reducing data means summarize, choose the main things, focus on the things that important, look for themes and patterns. In this study, data reduction was carried out when the researchers got data from Farmer Facilitator, Department of Agriculture Workers, Extension Workers, and Farmers about the obstacles in using the Among Tani Application, the efforts made to overcome these obstacles, as well as the expectations of the informants for the development of the application. Then it simplifies so that it leads to conclusions.

2. Data Display

In this research, the data were displayed using description and flowchart. At this stage the researcher describes the obstacles in using the Among Tani Application, the efforts made to overcome these obstacles, as well as the expectations of the informants for the development of the application to be easier to understood the meaning.

3. Verification

The conclusion drawing from this research was clear besause it was accompanied by evidences obtained in the field.

3. Results and Discussion

3.1. Existing Condition of Among Tani Application

The Among Tani application was created with the aim of providing services that facilitate the buying and selling process of horticultural products between farmers and buyers. Hopefully that the Among Tani Application could be integrated with all Regional Organizations. This application was divided into three applications, namely Among Tani, Among Kota, and Among Warga [11].In Among Tani Application there were two features that beneficial for farmers, they were compaint feature and discussion feature. In complaint feature, farmer just reported their complaint to the Extension Worker by sent documentations they had. And then they gave any feedbacks to the farmers less than two days. It been related with informant said:

"... I used the complaint feature to report diseases and about crop cultivation. So, later we will take a photo of the complaint and we will report it, after that tomorrow a team from extension worker will come to checked..." (Mr. Wijianto, 05 September 2021).

In discussion feature, farmers, extension workers, and farmer facilitator could discuss anything that they faced in farming. Whereas, the feature that most unbeneficial to used was offer feature. It was because this feature only gave the information about post-harvest that could not be related directly with marketing. It been related with informant said:

"...The rarest is an offer. I've opened an offer before but I don't think it's enough. The offer contained something like when it was ready to harvest, for example, rice was harvested in about a month, so it was just information. Meanwhile, it will be too long and the product will be easily damaged if it was vegetables. So yeah, I've never opened that feature..." (Mr Dendik, 20 September 2021)

Introducing process of new technologies was certainly not easy for some users to access Among Tani, especially for farmers. There are users who did not understand how to used application and all the functions of the Among Tani application features. It was because users still accustomed to received information manually. Changes in communication technology and information technology are one of the challenges for agricultural extension workers. There are farmers who do not adapt to changes in communication and information technology, and some are the opposite [12]. Some users stated that the use of the Among Tani application was not practical. There were any factors that affect unpractical utilization of application, they were not supported gadget, poor signal, and low capability human resources. [13] explained that one of the obstacles in the agricultural sector was the limited quality of human resources in agriculture. The low level of education of farmers causes their ability to absorb information and adopt technology was relatively limited, resulting in less production of quality products. So, the solution of these obstacles were to optimizing role of farmer facilitator to help farmer marketing through the innovations, such as "Protan" (Agricultural Promotion Product). It was the innovation program which help farmer to sold their product through social media, likes whatsapp, instagram, and shopee. So, farmer not lose the market.

"...It was very helpful, the farmers were happy with this. We saw that the offer feature of among tani was still not in line with farmers' expectations, finally we made a "Protan" through social media, through podcasts, etc., so facilitators helped from the design process, making stickers, to marketing..." (Mr. Andik, 05 September 2021)

3.2. Development Models of Among Tani Application

The distribution and supply chain of agricultural products were too long. Meanwhile, farmers were sometimes disadvantaged by prices that were too low. In this regard, it was hoped that initiatives and contributions from other institutions that will emerge to help these problems. Based on the results of interviews, the institution that was expected to be present in assisting the marketing of agricultural products were Department of Industry and Trade, banking, and also village own organization. Department of Industry and Trade, at least, this Department protected the farmers from the basic price. Beside that, this Department also helped the farmers to facilitate enterprise legacy.

"...Hopefully the facilitation will come from village own organization and also from Department of Industry and Trade. So far, if there were products that were too cheap/did not sell, the Department of Agriculture was angry. Meanwhile, the job of extension workers was not there, only on the technical side. How to managed ecosystems, the environment, improve quality in cultivation. Regarding the market, it should have been the duty Department of Industry and Trade, so they should have taken part in the market share." (Mr. Dulkamar, 07 September 2021)
"...surely village own organization and banking. Yesterday there was a submission to BNI hoped that farmers could cooperate by using e-money or OVO or Shope Pay. So facilitated online payments and product quality assurance.." (Mr. Adi, 07 September 2021)

Based on the interview above, it was hoped that later the features in the Among Tani application could be more easily accessed by users from farmers, extension workers, and facilitators. In addition, it was hoped that after the server repaired, promotional facilities will also been in the application. Some users hope that there will be socialization with smaller groups in order to develop this application into e-commerce platform. According to [14], the factors that affect in the development of farmer information include: farmer groups, farmers' learning intensity, the role of extension workers, the influence of outsiders, support for local wisdom, and characteristics of farmers.

"Nowadays, we discussed about the development of application. It will be changed into e-commerce platform. If there was no development, it was less effective. Promotional facilities and socialization for extension workers need to be reorganized, maybe the banners for each village, need to be improved too, so far we have only made it digitally.." (Mr. Andik, 05 September 2021)

4. Conclusions

It can be conclude that there were obstacles in using the Among Tani Application, they were limitation in human resources and access, especially in utilizing marketing features. So that farmers collaborate with farmer facilitator innovate to follow up on marketing that have disconnected from the application. Hopefully, that Among Tani Application changed into e-commerce platform and there was interference from the Department of Industry and Trading.

References

[1] Praditya, D. (2016). Pemberdayaan Petani oleh Komunitas Teknologi Informasi dan Komunikasi (TIK). Jurnal Penelitian Komunikasi 19 (1): 41-54.

[2] Amin, M et al. (2013). *The Effectiveness Of Cyber-Extension-Based Information Technology To Support Agricultural Activities In Kabupaten Donggala, Central Sulawesi Province, Indonesia. International Journal of Asian Social Science,* 2013, 3(4):882-889.

[3] Leeuwis, Cees. (2009). Komunikasi untuk Inovasi Pedesaan. Yogyakarta: Kanisius.

[4] Mulyandari, R.S., Sumardjo, Panjaitan, N.K., Lubis, D.P. (2010). Implementasi Cyber Extension dalam Komunikasi Inovasi Pertanian. Informatika Pertanian Volume 19, No. 2, IPB

[5] Mulyandari, Retno S.H. (2005). Alternatif Model Diseminasi Informasi Teknologi Pertanian Mendukung Pengembangan Pertanian Lahan Marginal. Prosiding Seminar Nasional Pemasyarakatan Inovasi Teknologi dalam Upaya Mempercepat Revitalisasi Pertanian dan Perdesaan di Lahan Marginal, Mataram, 30-31 Agustus 2005.

[6] Wahyuni.(2019).https://www.malangtimes.com/baca/41514/20190711/161000/pemkot-batu-dorong-petani-muda-kenali-aplikasi-among-tani. Diakses pada 24 Februari 2020.

[7] BPS dan Kemenpppa. (2017). Profil Perempuan Indonesia 2017. Jakarta: Kemenetrian Pemberdayaan Perempuan dan Perlindungan Anak.

[8] Aji, P.S., & Suparman. (2013). Pengaruh pembelajaran menggunakan macromedia flash 8 pokok bahasan internet pada mata pembelajaran TIK terhadap Prestas belajar siswa kelas XI IPA SMAN 6 Purworejo. Jurnal Pendidikan Teknik Informatika, 1.

[9] Setiawan, Iwan. Rasiska, Siska. Dkk. Korban "Tuhan-Tuhan" Digital: Garapan Masa Depan Penyuluhan Pembangunan 4.0. Indonesian Journal of Socio Economics, Volume 1, No 2, page 115-128 (2019).

[10] Sugiyono. (2014). Metode Penelitian Pendidikan Kuantitaif, Kualitatif, dan R&D. Bandung: Alfabeta

[11] Batukota.go.id, 2017. (Online) Tersedia di: https://www.batukota.go.id/ [Diakses 25 Oktober 2021]

[12] Ibrahim Jabal Tarik dan Mufriantie Fithri. 2021. Sumber Daya Manusia Sektor Pertanian, dalam berbagai perspektif. Malang: Psychology Forum bekerjasama dengan DPPs UMM.

[13] Wibowo, C. T., Sumardjo. Hafidhuddin, D., Agung, S. S. 2012. Pola Komunikasi Pada Pengembangan Kapasitas Kewirausahaan Petani Sayuran (Kasus Pendampingan Misi Teknik Taiwan Di Kabupaten Boyolali Dan Bogor). Jurnal Komunikasi Pembangunan. 10(1).

[14] Veronica, Helmi, Henmaidi, Arif, E.. 2018. Pengembangan Kapasitas Dan Kelembagaan Petani Kecil Di Kawasan Pertanian Melalui Pendekatan Manajemen Pengetahuan. Journal of Applied Agricultural Science and Technology.2(2).

Analysis of Organic Brown Sugar in Banyumas Regency

Dindy Darmawati Putri*, S Suyono, Irene Kartika Eka Wijayanti

Universitas Jenderal Soedirman, Purwokerto, Indonesia

*dindy.putri@unsoed.ac.id

Abstract. Banyumas Regency is one of the most potential organic brown sugar producers in Central Java. Organic brown sugar has a higher selling value than coconut sugar. The comparison of the price of coconut sugar and brown sugar which has a difference of Rp. 6,500.00 / kg with an average farmer production of 8 kg / day. The high price of brown sugar is very beneficial for coconut sugar farmers. This study aims to determine the income of the sugar craftsmen and determine the RC ratio. The research was conducted in two districts, namely Cilongok and Pekuncen districts, with 2 farmer groups, namely the Manggar Jaya farmer group and the Cikal Mas farmer group. The basic method used in this research is descriptive analysis. Data analysis using cost income analysis and RC ratio. The results showed that the craftsmen revenue Rp106,087.00 per day and the value of the RC ratio was 1.40 indicating that the business was feasible to run.

Keywords: business analysis, RC ratio, brown sugar ,agroindustry

1. Introduction

Banyumas Regency is one of the potential producers of brown sugar in Central Java. According to 2017 statistical data obtained from the DisperindagkopBanyumas Regency, the production of brown sugar in Banyumas reaches 35,562 tons per year with 7,863 business units, and as much as 25,819 tons of brown sugar are exported to various countries. In 2015, the average consumption of coconut sugar per capita per week reached 0.136 ounces and consumption per capita per year reached 0.709 kg. Based on data from the Directorate of Small and Medium Trade and Domestic Products, the Ministry of Trade, the market demand for Indonesian brown sugar products reaches 400 tons consisting of domestic and foreign demand. Domestic demand alone reaches 20 tons per month, while national production capacity is only in the range of 5 – 10 tons per month. Thus, although it is still export-oriented because it is considered more profitable, the domestic market is still growing.

There are 2000 brown sugar craftsmen who have obtained organic certification from the organic brown sugar certification agency Control Union Certification (CUC) which includes organic standards from USDA (United States), NOP (Europe) and JAS (Japan). Organic certification shows that Banyumas organic brown sugar products have been able to penetrate foreign markets (exports) [1]. Generally, the main consumers of brown sugar products, especially North America and Northwest Europe, are the centers of consumption. Viewed in more detail, countries with high consumption of coconut sugar are the United States, Germany, the Netherlands, Ireland, Australia, Belgium. England, Austria and Scandinavia.

The development of agro-industry in rural areas is one of the government's programs to support the improvement of the quality of Indonesia's exports, from commodity products to products that have certification and branding. The development of agro-industry in rural areas is primarily aimed at increasing the added value of agricultural commodities, expanding employment opportunities, increasing export volume and encouraging rural economic growth. Utilization of local food raw materials in agro-industry is expected to increase the diversity and availability of food products. The problems faced in the brown sugar agroindustry are:
1. How is the income and efficiency of the brown sugar agroindustry
2. What is the flow pattern and supply chain mechanism of brown sugar agroindustry
From these problems, this research aims to:
1. analyze the income and efficiency of the brown sugar agroindustry
2. analyze the flow pattern and supply chain of brown sugar agroindustry in Banyumas Regency

2. Research Method

The basic method used in this research is descriptive analysis, which is a research method to examine a group of people, an object, a set of conditions, a thought or a class of events in the present or at least a period of time that is still affordable in the respondent's memory. is sequential explanatory which combines research techniques, methods, quantitative and qualitative approaches in one study.
The study lasted for 8 months, namely in March-November 2020. The study was conducted in Banyumas Regency. The location selection was carried out purposively, namely in Pekuncen and Cilongok Districts, Banyumas Regency by considering that the location is the largest brown sugaragro-industry center in Banyumas Regency. The method of data collection is done by survey technique. In the survey research method, information is obtained through direct interviews with respondents and observations. Data analysis using Cost and Revenue Analysis and RC ratio.[2]

3. RESULTS AND DISCUSSION

Cost, Revenue, Revenue and Profit Analysis Cost analysis is used to determine and calculate the total costs incurred during the production of brown sugar. The costs analyzed include fixed costs, variable costs, total costs as well as revenues and profits received by the brown sugar business owner in the calculation for one time of production. Cost analysis can make it easier for business owners to find out the amount of costs used in the production process, the amount of revenue and profits received and can find out the business efficiency of the brown sugar agroindustry.

Revenue. Revenue is the result of multiplying the number of products with the selling price per unit of product. The amount of brown sugar produced per day is 5.3 kilograms with a selling price of Rp. 20,000 per kilogram. The average revenue earned by craftsmen is Rp.106,087.00 per day.This revenue is slightly higher than the study [2]on brown sugar in KulonProgo, Yogyakarta, which is Rp. 95,450/day.

Profit. According to[3], income is the difference between total revenue and total costs (variable costs and fixed costs). The average income obtained from the brown sugar agroindustry in Banyumas Regency can be seen in Table 1.

Table 1. Average Income of Brown sugar Agroindustry in Banyumas Regency per day

Description	Average (Rp/day)
Revenue	106.086,95
Variable Cost	14.961,00
Fixed Cost	60.566,14

Total Cost Total	75.527,14
Profit	**30.559,81**

Source: Primary data processed, 2020

Table 1. shows that the brown sugar agroindustry in Banyumas Regency earns anprofit of Rp. 30,559.81 per day. This profit is obtained from the total revenue of Rp.106,086.95 per day minus the total cost of Rp.75,527.14 per day.

Business Efficiency Analysis Efficiency is defined as the effort to use the smallest input to get the maximum production. Efficiency can be known by calculating the Revenue of Cost Ratio (R/C). R/C is the ratio between total revenue and total cost 2003)[4].
Mathematically :

$$RC\ ratio = \frac{Total\ Revenue}{Total\ Cost}$$

If R/C > 1	then the business is efficient (profitable) to be cultivated, meaning that the input used is appropriate.
If R/C < 1	then the business is not efficient (not profitable) to be cultivated, to achieve efficiency, the input needs to be added.
If R/C = 1	then the business is break-even, i.e. the business provides the same amount of revenue as the amount spent.

R/C analysis is used to determine the value of the comparison between revenues and costs, so that it can be seen the amount of revenue obtained from each unit of rupiah costs incurred. The amount of R/C value in the brown sugar agroindustry in Banyumas Regency per day shows more than one, namely 1.40 or it can be said that the revenue obtained by the brown sugar agroindustry in Banyumas Regency is greater than the total costs incurred in the production of brown sugar, then the agro-industry business brown sugar is efficient (profitable) to cultivate, meaning that the input used is appropriate. The R/C value of 1.40 means that the brown sugaragro-industry in Banyumas Regency will receive an income of Rp. 1.40 from every Rp. 1.00 of the costs incurred. [5] stated that the greater the R/C value, the greater the profit from the business.
Based on previous research conducted by [6]with the title Comparative Analysis of Coconut Farmers' Income from Brown sugar Business with Printed Sugar in Hargobojo Village, Bagelen District, Purworejo Regency, it is efficient and able to generate profits because the brown sugar business in Hargobojo Village has a higher R/C value. from 1 which is 1.25. This shows that the brown sugar business in Hargobojo Village is more efficient than the brown sugaragro-industry in Banyumas Regency because it has a higher R/C value. The costs incurred in the brown sugar business in Hargobojo Village are lower than the brown sugar agroindustry in Banyumas Regency, and different selling prices will affect the R/C value obtained.
However, these results are slightly lower than the research of[2] with RC rati0 1.42 and[7]inCiamis Regency with RC ratio 1.43

The Supply Chain of Brown Sugar Agroindustry Supply chain of brown sugar agroindustry. There are several supply chain patterns for gulasemut in Banyumas Regency as follows:

1. Supply chain pattern 1

In this pattern, producers (penderes) sell palm sugar products directly to supplier traders to exporters (suppliers).

Craftsmen -----→ Suppliers -----→Exporters

2. Supply chain pattern 2

This pattern is a marketing activity that is quite a lot carried out by producers, namely producers sell to collectors traders around their homes, then collectors sell directly to supplier traders to exporters (suppliers).

Craftsmen ----------→Collectors------→ Suppliers --------→Exporters

3. Supply chain pattern 3

This supply chain pattern is the pattern most used by producers, namely producers sell to collectors, then collectors sell back to wholesalers, then wholesalers sell to supplier traders, and finally supplier traders sell back to exporters.

Craftsman -----→ Collector-----→Wholesaler ----→Supplier ----→Exporter

4. Supply chain pattern 4

In this supply chain pattern, craftsmen sell directly to wholesalers, then sell them again to supplier traders and finally to exporters.

Craftsman ----→Wholesaler ----→ Supplier ----→ Exporter

5. Supply chain pattern 5

In this pattern, consumers buy directly from producers (penderes), this consumer segment wants to get brown sugar products that are cheap, even though the water content is still quite high, which is around (8-10) percent. Consumers in this segment need to dry the brown sugar they buy for about 2-3 hours in the hot sun so that the water content reaches (3-4) percent, with the aim that brown sugar products can last for at least 6 months when packaged and stored in free air conditions and dry place.

4. Conclssion and Suggestion

A. Conclusion

1. Household-scale agro-industry revenue in Banyumas Regency is IDR 106,087.00 per day with a total production of 5.3 kilograms of ant sugar per day. income of IDR 30,559.81 per day. The value of R/C in the brown sugar agroindustry in Banyumas Regency per day shows more than one, namely 1.40.
2. There are 5 patterns of supply chains in Banyumas Regency consist of craftment, collector, wholesaler, supplier and exporter

B. Suggestion

1. To improve the performance of agro-industry and management, it can be done by increasing supply chain activities through partnership strategies and good relationships with customers or consumers
2. Income from craftsmen can be increased with a good marketing management system so that customers will increase
3. Business efficiency can be increased by increasing the number of crushed coconuts, namely by expanding the area of coconut plantations in Banyumas Regency

Acknowledgment

This research was funded by the JenderalSoedirman University, BLU Fund through the 2020 Competency Improvement Research research scheme

References

[1] E. Khikmawati, "Kontribusi Agroindustri gula semut Terhadap Ekonomi Rumah Tangga Pengrajin gula semut di Desa Rancamaya Kecamatan Cilongok Kabupaten Banyumas.," Universitas Muhammadiyah , Purwokerto, 2015.

[2] D. Putri, "Studi Komaparatif Pendapatan Biaya, dan Kelayakan Usaha Agroindustri Gula Semut pada Setiap Pelaku Rantai Pemasaran," *SEPA,* vol. 17, no. 1 September 2020, pp. 65-71, 2020.

[3] B. N and H. M. T. Georgius, "Analisis Kelayakan Ekonomi Usahatani Kedelei di Desa Kebun Agung Kabupaten Grobogan," *Proceeding Konser Karya Ilmiah Nasional,* no. 2, pp. 2460-5506, 2016.

[4] Soekartawi, .Prinsip Ekonomi Pertanian, Jakarta: Rajawali Press, 2003.

[5] Munawir, Analisis Laporan Keuangan Edisi 4, Yogyakarta: Liberty, 2010.

[6] N. Fauztina, "Analisis Komparasi Pendapatan Petani Kelapa Dari Usaha Gula Semut Dengan Gula Cetak Di Desa Hargorojo, Kecamatan Bagelen, Kabupaten Purworejo.," *Jurnal Ekonomi Pertanian dan Agribisnis,* vol. 3, no. 1, pp. 84-95, 2019.

[7] I. S. N. Z. Hadwa, "Analisis Usaha dan Nilai Tambah Gula Semut Studi Kasus di Desa Sidamulih, Kecamatan Pamarican Kabupaten Ciamis," *Agroinfo Galuh,* vol. 3, no. 2, pp. 220-225, 2017.

[8] K. Soeratiyah, Ilmu Usahatani, Penebar Swadaya: Jakarta, 2015.

IOP Conf. Series: Earth and Environmental Science 1131 (2023) 012011 doi:10.1088/1755-1315/1131/1/012011

Survivability and Benefit Evaluation of Streptomyces sp. and Trichoderma sp. as Active Ingredients of Biopesticides and Soil Fertility Enhancer in Shallot Fields at Wates Village Tulungagung

Rachmanita D Prastiti, Arga D Indrawan, Penta Suryaminarsih*, Tri Mujoko, Bakti W Widjajani

UPN "Veteran" Jawa Timur, Surabaya, Indonesia

*penta_s@upnjatim.ac.id

Abstract. The purpose of research on the Farmer Empowerment Program in Wates Village is to increase knowledge in the field of sustainable agriculture and transfer of technology for the application of biopesticides containing the Biological Control Agent (APH) *Streptomyces* sp., *Trichoderma* sp. combined with shallot cultivation technology in the village of Wates Tulungagung. Innovative methods and Participatory Action Research (PAR) are used to encourage transformative action, by involving community members as PAR implementers in describing the types of plants being treated and applying them during the Covid-19 pandemic in the Wates village, Tulungagung District. The results of the evaluation and monitoring of the use of APH as a biopesticide and fungicide to control moler disease caused a decrease in the population of microorganisms, the population of *Trichoderma* sp. was absent (o) but *Streptomyces* sp. increase. Giving APH proved that microorganisms act as decomposers that increase the availability of N and P nutrients through their ability to accelerate the decomposition of organic matter given at the beginning of planting. The K nutrient content in the soil before and after treatment did not differ.

1. Introduction

Wates Village is located in a lowland area. The dry land in wates village used for vegetable cultivation consisted of 11 ha of chili, 7 ha of tomatoes, 6 ha of long beans and 17 ha of eggplant. The dry land has an irrigation system in the form of 14 ha of semi-technical irrigation and 73.11 ha of rain fed [2]. The dependence of farmers on chemical fertilizers and pesticides is very high, so they do not pay attention to food safety and the environment.

Based on the evaluation results from joint learning activities and the knowledge of farmers about farming methods that are carried out every day, farmers already know about organic fertilizers and environmentally friendly pesticides but do not understand and have not implemented them seriously. In general, farmers in this village grow tobacco, chili, tomatoes, eggplant. However, during the Covid-19 pandemic, farmers suffered a lot of losses because production costs were quite high, especially labor and the prices of chili and other horticultural crops decreased drastically. During the Covid-19 Pandemic, farmers began to switch to shallot plants which were more promising for selling value. Many farmers have turned to shallot plants even though they have never planted these plants.

The Innovation Method is a method of community empowerment that can change the culture or habits of the community with a new culture or technology [3]. Participatory Action Research (PAR) is a research method that is carried out in a participatory manner among members of the community in a lower-level community that encourages transformative actions. [7]

The use of Microorganisms in sustainable agriculture as one of the nano technology innovations for the safe, environmentally friendly and sustainable control for plant-disturbing organism [14]. *Streptomyces* sp., as a biological agent is also a microbial degrading carbon in plant residues so that it can be available to plants and also as a decomposer of recalcitrant protein in proteolysis [9]. *Trichoderma* sp. is a biopesticide and a biofertilizer, it improves soil conditions and has the ability as a biological agent to pathogenic fungal with various mechanisms such as parasitism, competition and antibiosis [8, 19].

The results of research by Suryaminarsih, Kusriningrum, Surtiningsih and Ni'matuzahroh [21] showed that the combination of biological agents *S. griseorubens* f.sp. capsicum, *G. virens* and *T. harzianum* are compatible and can inhibit the development of *F. oxysporum* under in vitro conditions.) and have the ability to inhibit the development of the severity of fusarium wilt disease caused by *F. oxysporum* f.sp. lycopersici and increase production. Utilization of *Streptomyces* sp., *Trichoderma* sp. as decomposers and biological agents by farmers in Pare village can reduce the presence of pests and diseases and increase plant growth almost the same as what they usually do [20].

2. Materials and methods

2.1 Tools and materials

Biological Control Agent (BCA) *Streptomyces* sp. and *Trichoderma* sp. is the result of technological development resulting from research that has been carried out sequentially and consistently since 2012. This composition of BCA has been registered as a simple patent with application letter number S00202005990. The ECG media sterilizer was assembled by TTG UPN "Veteran" Jawa Timur. The BCA propagation tool, the Fermentor, was assembled by Pusat Pelatihan Agensia Hayati (PPAH) Pasuruan.

2.2 Production and application of biopesticides containing Streptomyces sp., Trichoderma sp.

Streptomyces sp. isolates as many as 3 petri dishes with a diameter of 9 cm and *Trichoderma* sp. isolates as many as 1 petri dishes were put into 20 L of sterile Potato Sugar Extract (EKG) medium, then fermented for 10 days using a simple fermenter.

The propagation of biological agents can be done by farmers themselves so that farmers can use these biological agents to make organic fertilizers independently. The mechanism of propagation and application of biological agents is carried out as shown in the scheme below (Figure 1)

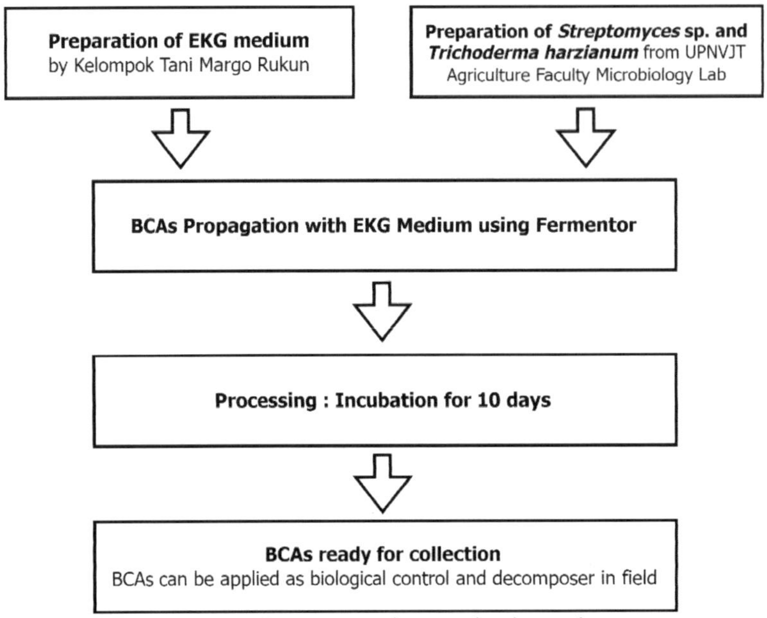

Figure 1. BCA Propagation mechanism scheme

Harvesting of biological agents (BCAs) propagated in the fermenter must have an odor that can indicate good results without contamination, which will be applied to demonstration plots that are treated with synthetic fertilizers and organic fertilizers (manure, compost or bokashi). Applications for demonstration plots and assistance for application of biological agents to plants and monitoring of the sustainability of sustainable crop management are carried out on land that has been prepared by the group leader before planting, on shallot seeds and 4 times during the planting period. Monitoring and evaluation of the results of the demonstration plot is carried out by the farmer groups themselves by making notes on cultivation needs and production results. Evaluation and monitoring of the content of microorganisms and NPK was carried out at the Laboratory of UPN "Veteran" Jawa Timur after harvest. Evaluation of the number of microorganisms and the nutrient content of NPK was carried out by taking soil samples of Fallow soil after horticultural planting (TB) and shallot land with BCA Application technology along with Moler/Fusarium Disease Control Fungicides (TPHF).

3. Results and discussion

3.1. Survivability Evaluation of Streptomyces and Trichoderma in the Agricultural Land

Management and treatment of agricultural land can directly affect the presence and number of soil microorganisms in it. Figure 2 shows the results of the BCA application in the two types of land management showing differences in the types and numbers of microorganisms found.

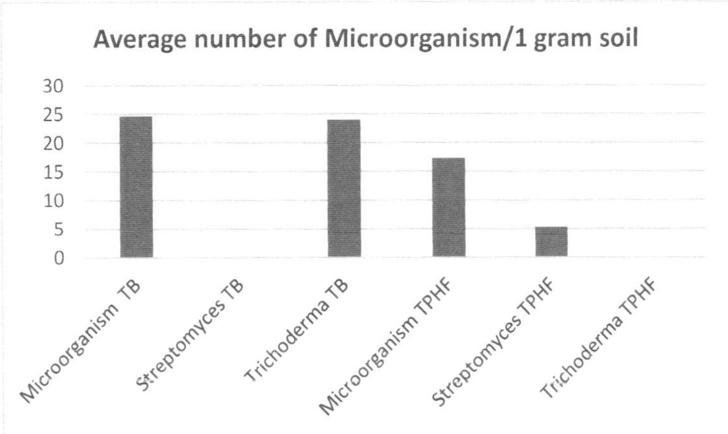

Figure 2. Average number of microorganism/1 gram soil

Figure 2 shows that in the TB treatment area, *Trichoderma* (TB) was found as much as 24 x 10/gram soil and the total number of TB soil microorganisms (Microorganism TB) was 24.67 x 10/gram soil, but no *Streptomyces* (TB) colonies were found. Meanwhile, on land with TPHF treatment, *Streptomyces* was found as much as 5.33 x 10/gram soil and the total number of TPHF soil microorganisms (Microorganism TPHF) was 17.33 x 10/gram, but *Trichoderma* (TPHF) was not found. Overall the number of TB microorganisms is higher than the number of microorganisms, this is thought to be influenced by the way of management carried out. The overall population of soil microorganisms in the low TPHF and indicated by the absence of *Trichoderma* as BCA from the fungal group in the soil was thought to be caused by the use of fungicides. Ons et al. [14] stated that there are several biological agents such as those from the *Trichoderma* genus which are susceptible to chemicals, especially fungicides, so that BCA applications are not compatible when used together with fungicides. Therefore, this incompatibility can reduce the effectiveness of Trichoderma in the field.

Specifically, the population of BCA *Streptomyces* on land treated with fungicides (TPHF) was higher than that on fallow soil (TB). The persistence of *Streptomyces* in TPHF fields is thought to be due to its tolerance to fungicide active ingredients. Lee et al. [10] found several fungicide-tolerant *Streptomyces* strains isolated from fields with intensive fungicide use. The absence of *Streptomyces* in the fallow land (TB) is thought to be due to its slow growth, the time lag between one planting period and the next, causing the absence of host plants in the ecosystem during the fallow period. Lopez-Mondejar et al. [11] explained that some species of Actinobacteria have the ability to grow slowly, so they need an immediately available source of nutrients such as exudates produced by plant roots so that *Streptomyces* generally live and interact around plant roots. This is in accordance with the statement of Sousa and Olivares [18] and Olanrewaju and Babalola [13] that some *Streptomyces* live in the rhizosphere and others live endophytes by colonizing the root tissue of host plants in a mutualism symbiosis as BCA and PGPS (Plant Growth-Promoting *Streptomyces*) living with colonize plant tissues, in addition Worsley et al. [23] stated that *Streptomyces* obtained plant root exudates as a source of nutrition when colonizing roots.

Keberadaan tanaman umumnya juga mempengaruhi kondisi lingkungan disekitarnya yang disebut mikroklimat. Ketiadaan tanaman budidaya pada masa bera diduga dapat menyebabkan tanah dapat terpapar sinar matahari secara langsung dan diduga mempengaruhi iklim mikro khususnya suhu dan kelembapan tanah. Suhu tanah yang tinggi dapat mengurangi kelembapan sehingga menghambat pertumbuhan *Streptomyces*, sehingga pada tanah bera(TB) tersebut tidak ditemukan populasi *Streptomyces* yang bertahan. Hal ini dijelaskan oleh Suryaminarsih et al. [22] bahwa dalam pengujian efektivitas dan pertumbuhan multiantagonis pada temperature berbeda, *Streptomyces* tidak dapat tumbuh dengan baik lingkungan dengan suhu lebih dari 30°C. Berdasarkan informasi tersebut, diperlukan lingkungan yang mendukung supaya efektivitas streptomyces sebagai BCA tetap optimal. Berdasarkan informasi tersebut, diperlukan lingkungan yang mendukung dan sesuai supaya efektivitas streptomyces sebagai BCA tetap optimal.
The presence of plants generally also affects the surrounding environmental conditions, which is called the microclimate. The absence of cultivated plants during the fallow period is thought to cause the soil to be exposed to direct sunlight and is thought to affect the microclimate, especially soil temperature and humidity. High soil temperatures can reduce humidity thereby inhibiting the growth of *Streptomyces*, so that in the fallow soil (TB), no surviving *Streptomyces* population is found. This is explained by Suryaminarsih et al. [20] that in testing the effectiveness and growth rate of multi-antagonists at different temperatures, *Streptomyces* could not grow well in an environment with a temperature of more than 30°C. Based on this information, a supportive and suitable environment is needed so that the effectiveness of streptomyces as a BCA remains optimal.

3.2. Impact of Streptomyces and Trichoderma on Soil Fertility

In addition to their role as BCA, *Streptomyces* and *Trichoderma* have important functions for plant growth by increasing soil fertility. The test results on several indicators including soil pH and nutrient content before and after the application of *Streptomyces* and Trichoderma are presented in Table 1.

Table 1. Soil nutrition content

No.	Parameter	Sample			Criteria
		TLK	TB	TPHF	
1	pH	6.1	6.0	6.1	Slightly Acidic
2	C	1.19 %	1.27%	2.03 %	Low
3	N	0.09 %	0.12%	0.18%	Low
4	P	22 ppm	23 ppm	27 ppm	High
5	K	0.53 me/100g	0.55 me/100g	0.55 me/100g	Medium

Table 1 shows that the pH condition of the soil before the application of *Streptomyces* and *Trichoderma* (TLK) was 6.1, after application there was a decrease in the pH of the fallow soil (TB) to 6.0 but there was no change in the pH value of the TPHF land so that the pH value remained at 6.1. The change in pH value to slightly more acidic in TB land was thought to be due to the activity of decomposition of organic matter by microorganisms, this is in accordance with the statement by Fondriest Environmental [5] that the decomposition process is one of the causes of the decrease in pH. In addition, according to Pelagio-Flores et al. [16] *Trichoderma* activity can induce soil acidity to form an optimal environment for its growth and to stimulate plants (phytostimulation).
Based on Table 1, there was an increase in C levels from the previous 1.19%, after *Streptomyces* and *Trichoderma* were applied to 1.27% in TB fields and 2.03% in TPHF fields. In addition to the addition of C levels, the administration of *Streptomyces* and *Trichoderma* increased the macronutrients N, P, and K, respectively, from 0.09%; 22 ppm; and 0.53 me/100 g on TLK land, to 0.12%; 23 ppm; and 0.55 me/100 g in TB field and 0.18%; 27 ppm; and 0.55 me/100 g on TPHF fields. The addition of these nutrient levels is thought to be due to the process of decomposition of soil organic matter into nutrients

that can be absorbed by plants. According to Nduka et al. [12] *Streptomyces* and *Trichoderma* are beneficial soil microorganisms that act as decomposers of organic matter and Plant Growth-Promoters as well as biological agents. Specifically, *Trichoderma* can increase soil fertility by increasing the availability of macronutrients N, P, and K as well as other micronutrients through the process of decomposition of organic matter and nutrient fixation [6].
The role of *Streptomyces* as decomposers is mentioned by Lopez-Mondejar et al. [11] that *Streptomyces* as part of Actinobacteria have a role in the C carbon cycle in the ecosystem. *Streptomyces* has 10 enzymes that can be used in the degradation process of polysaccharides and lignin in lignocellulose in plants and fungi [4]. Based on the decomposition ability, it is suspected that *Streptomyces* contributed to the increase in element C and other nutrients which were also released during the decomposition process in TPHF land.
Based on the discussion, it is known that *Streptomyces* and *Trichoderma* have dual benefits which can also act as an increase in soil fertility as well as important information to be used in determining good management methods to support BCA sustainability on land. To support the sustainability and sustainability of BCA, managers can pay attention to soil and environmental conditions, BCA's compatibility with cultivated plants, availability of organic materials, use of natural pesticides that are compatible with BCA, and seek chemical pesticides as a last resort in the process of controlling shallot pests [17].

4. Conclusion

The conclusion from the results of socialization, training, demonstration plots of shallot planting and monitoring evaluation shows that with the PAR method the community is able to cultivate horticultural crops, but has not been able to utilize organic fertilizers optimally and sustainably, Chemical Pesticide Control is still high. Farmer groups have been able to make and reproduce APH and apply it to shallot plants. Monitoring and evaluation of the implementation of BCA carried out with farming communities resulted in information that *Streptomyces* and *Trichoderma* increase the availability of organic matter or C and increase the availability of macronutrients N, P, and K but the impact and yield may differ depending on the applied land management pattern.
Suggestions for sustainable assistance to lead a sustainable agricultural system are still needed where with this community service program it is hoped that the existence of a Land Resource Laboratory, a Plant Health Laboratory of the Agrotechnology Study Program of the Faculty of Agriculture can continue to develop, reciprocally and sustainably. Agricultural Extension Officers from the Governmental Agency in Food Security Service and Subvillage Government need to intervene intensely to maintain the existence and empowerment of farmer groups towards a sustainable agricultural system.

References

1. Fernandes TAR, da Silveira WB, Passos FML et al.. (2014) Laccases from Actinobacteria. What we have and what to expect. Adv Microbiol 2014;4:285–96.
2. Fondriest Environmental, Inc. (2013). pH of Water. Environmental Measurement Systems. Retrieved June 11, 2022, from https://www.fondriest.com/environmental-measurements/parameters/water-quality/ph/
3. Halifu, S., Deng, X., Song, X., & Song, R. (2019). Effects of Two Trichoderma Strains on Plant Growth, Rhizosphere Soil Nutrients, and Fungal Community of Pinus sylvestris var. mongolica Annual Seedlings. Forests, 10(9), 758. https://doi.org/10.3390/f10090758
4. Indrawan, A. D., Suryaminarsih, P., & Mujoko, T. (2021). Prospect of utilization of microorganisms Streptomyces sp. and Trichoderma sp. in supporting sustainable agriculture in the age of modern agriculture. Sains dan Teknologi Pertanian Modern. NST Proceedings. pages 32-38. doi:10.11594/nstp.2021.1506

5. Lange, L., Huang, Y. & Busk, P.K. 2016. Microbial decomposition of keratin in nature—a new hypothesis of industrial relevance. Appl Microbiol Biotechnol 100, 2083–2096 (2016). https://doi.org/10.1007/s00253-015-7262-1
6. Lee, J.H., Min, G.Y., Jeon, C., Choi, S.M., Shim, G.Y., & Kwak, Y. 2015. Investigation of Fungicide Response of Streptomyces spp. Isolated from Rhizosphere in Zoysiagrass. Biosciences, 8. https://doi.org/10.3389/fmolb.2021.686110
7. López-Mondéjar, R., Brabcová, V., ŠTursová, M., Davidová, A., Jansa, J., Cajthaml, T., & Baldrian, P. (2018). Decomposer food web in a deciduous forest shows high share of generalist microorganisms and importance of microbial biomass recycling. The ISME Journal, 12(7), 1768–1778. https://doi.org/10.1038/s41396-018-0084-2
8. Nduka, B. A., Oduwaye, O. F., & Adewale, D. B. (2017). Potential of Streptomyces sp. and Trichoderma sp. as compost microbiota for coffee husk. African Journal of Microbiology Research, 11(14), 560–567. https://doi.org/10.5897/ajmr2017.8476
9. Olanrewaju, O. S., & Babalola, O. O. 2019. Streptomyces: implications and interactions in plant growth promotion. Applied microbiology and biotechnology, 103(3), 1179–1188. https://doi.org/10.1007/s00253-018-09577-y
10. Ons, L., Bylemans, D., Thevissen, K., & Cammue, B. 2020. Combining Biocontrol Agents with Chemical Fungicides for Integrated Plant Fungal Disease Control. Microorganisms, 8(12), 1930. https://doi.org/10.3390/microorganisms8121930
11. Pasandaran, E., M. Syakir RH., dan M. Prama Yufdy. 2017. Menuju Pertanian Berkelanjutan. Kementrian Pertanian, Badan Penelitian Pertanian.
12. Pelagio-Flores, R., Esparza-Reynoso, S., Garnica-Vergara, A., López-Bucio, J., & Herrera-Estrella, A. (2017). Trichoderma-Induced Acidification Is an Early Trigger for Changes in Arabidopsis Root Growth and Determines Fungal Phytostimulation. Frontiers in Plant Science, 8. https://doi.org/10.3389/fpls.2017.00822
13. Scialabba, N., Gomez, I., & Thivant, L. 2015. Training Manual for Organic Agriculture [E-book]. Food and Agriculture Organization of the United Nations. https://www.fao.org/fileadmin/templates/nr/sustainability_pathways/docs/Compilation_techniques_organic_agriculture_rev.pdf
14. Sousa JAJ, Olivares FL. 2016. Plant growth promotion by Streptomycetes: ecophysiology, mechanisms and applications. Chem Biol Technol Agric. 2016;3(1):24.
15. Suryaminarsih, P., Harijani, W. S., Syafriani, E., Rahmadhini, N., & Hidayat, R. 2019. Aplikasi Streptomyces sp. Sebagai Agen Hayati Pengendali Lalat Buah (Bactrocera sp.) dan Plant Growth Promoting Bacteria (PGPB) pada Tanaman Tomat dan Cabai. AGRIUM: Jurnal Ilmu Pertanian, 22(1), 62-69.
16. Suryaminarsih, P., Haryani, W. S., & Mujoko, T. 2017. Pemberdayaan Kelompok Tani dalam Meningkatkan Produksi Tomat dengan Pemanfaatan Agensia Hayati Streptomyces s p., & Trichoderma sp. Berkala Ilmiah AGRIDEVINA, 5(2).
17. Suryaminarsih, P., Kusriningrum, N. M., & Surtiningsih, T. 2015. Antagonistic Compatibility of Streptomyces griseurubens, Gliocladium virens and Trichoderma harzianum againts Fusarium oxysporum couse of Tomato Wilt Deseases. Int. J. Plant Soil Sci, 5, 82-89.
18. Suryaminarsih, P., Megasari, D., & Mujoko, T. 2020. The effect temperature pressure on multiantagonist Streptomyces sp., Trichoderma sp. biological control of Fusarium oxysporum wilt pathogens. Seminar Nasional Magister Agroteknologi Fakultas Pertanian UPN "Veteran" Jawa Timur. NST Proceedings. pages 85-91.doi:10.11594/nstp.2020.0609
19. Worsley, S. F., Macey, M. C., Prudence, S. M. M., Wilkinson, B., Murrell, J. C., & Hutchings, M. I. 2021. Investigating the Role of Root Exudates in Recruiting Streptomyces Bacteria to the Arabidopsis thaliana Microbiome. Frontiers in Molecular Biosciences, 8. https://doi.org/10.3389/fmolb.2021.686110

IOP Conf. Series: Earth and Environmental Science 1131 (2023) 012012 doi:10.1088/1755-1315/1131/1/012012

Phytochemical Compounds and Antibacterial Activity to Escherichia coli of Green Macro Algae

Ketut Srie Marhaeni Julyasih[1*] and Arika Purnawati[2]

[1]Universitas Pendidikan Ganesha, Singaraja Bali, Indonesia
[2]Agrotechnology, Agriculture Faculty UPN Veteran Jawa Timur, Surabaya, Indonesia

*smjulyasih@gmail.com

Abstract. Escherichia coli is a bacteria found in the large intestine of humans as a normal flora. It's unique because it can cause primary infection in the intestine such as diarrhea in children. Diarrhea sufferers use drugs derived from chemicals and herbal plants. Furthermore seaweed as the marine biological resources it is important to examine to be an alternative treatment for diarrhea. An antibacterial alternative by development the use of seaweed. needs to be examine. Based on the description above, phytochemical compound of several types of green macro algae and its potential effect as antibacterial activity to inhibit the growth of E.coli bacteria it is necessary to examine, The type of seaweeds that use in this study are Ulva reticulata, Ulva lactuca and Caulerpa racemosa. The research results showed that phytochemcal compound of U. reticulata are terpenoid, phenolic , alkaloid. In U. lactuca are steroid, saponin, phenolic and alkaloid. Phytochemcal compound of C. racemosa are phenolic, flavonoid, alkaloid, and terpenoid. Type of green macro algae U. reticulata had the ability to inhibit the bacteria E.coli growth with a inhibition zone diameter about 6.0 mm U. lactuca about 10 mm, and C. racemosa about 12.0 mm.

1. Introduction

Escherichia coli is a bacteria found in the large intestine of humans as a normal flora. It's unique because it can cause primary infection in the intestine such as diarrhea in children. Diarrhea sufferers use drugs derived from chemicals and herbal plants, but there are still limited studies that make one of the marine biological resources such as seaweed to be an alternative treatment for diarrhea [1]

Treatment of diarrhea until now is done with causative treatment, namely the germs that cause it to be turned off with antibacterial ingredients. Antibacterial compounds must be effective in controlling bacterial growth and the problem of resistance to materials used, especially bacteria that harm humans. The use of seaweed as in inhibiting the growth of E.coli bacteria it is important to be developed and realized.

Seaweed contains many chemical compounds as primary metabolites called hydrocolloids. Various industrial materials have been used hydrocolloids such as gelatin, carrageenan, alginates, and so on. In addition to primary metabolite products, secondary metabolite products from seaweed have begun to be widely studied and developed. Bioactive compounds are one of the secondary metabolites that have a lot of potential to be developed and researched as antimicrobials such as antibacterial, antifungal, antiviral, antioxidant and so on. [2].

Based on the description above, this research purpose is to analyze compound of Ulva reticulata, Ulva lactuca , and Caulerpa racemose phytochemical, and its potential effect as antibacterial activity to Escherichia coli growth.

2. Material and Methods.

2.1 Material

The materials that used in this research are three (3) kind of seaweeds that collected from Sanur Beach Bali like Ulva lactuca, Ulva reticulata, Caulerpa racemose, media NA for testing of seaweeds extract against bacteria E.coli, materials for phytochemical content of seaweeds

2.2 Methods

Studies was conducted in Laboratory Genetic Resource University of Udayana

Preparation of Seaweeds Extract. Samples macro alga washed with water, wind dried for 14 days in the open air without being exposed to direct sunlight. Dried seaweed that has been crushed with a blender and filtered with maceration for 24 hours in 96% ethanol solvent, seaweed has been macerated with Whatman filter paper No. 41. The result of seaweed extraction with ethanol solvent was then evaporated using a vacuum evaporator at a temperature of 49-50 ^{0}C until all the solvent evaporated, resulting in a crude extract of seaweed.

Analysis of Phytochemical Content of Several Types of Green Macro Algae Analysis of the phytochemical content of green macro algae was carried out by testing the content of alkaloids, terpenoids, steroids, phenolics, and flavonoids. Qualitative test of the phytochemical content of seaweed in 96% ethanol extract was carried out with chemical reagents to identify terpenoids, steroids, flavonoids, phenolics, and alkaloids. Phytochemical screening or phytochemical tests were carried out to see the bioactive components in the crude extract of seaweed. A total of 0.05 g of sample was reacted with each reagent to determine the bioactive content qualitatively. The appearance of pink, magenta and orange colours indicate contain of flavonoid compounds. The presence of a precipitate or turbidity indicates an alkaloid compound, a brownish red colour or a brownish pink ring indicates a terpene compound [3]

Testing of Green Macro Algae Extract Against Escherichia coli Bacteria. Before testing the antibacterial activity, sterilization of the equipment and materials to be used in an autoclave for 15 minutes at a temperature of 121 ^{0}C. NA seeding media was made by dissolving 23 grams of NA into 1 L of distilled water then heated until dissolved. Seaweed extract was put into a petri dish and then poured with 20 ml NA media which was still liquid. The media was filled with 40μl suspension of test bacterial spores. Media was made a hole with bore cork Ø 5 mm and seaweed extract was put in to a hole. Bacteria were obtained from the Bacteriology and Mycology Laboratory of the Faculty of Veterinary, University of Udayana. The parameters test of the inhibition zone (mm) of each green macro algae extract were observed using a ruler or calipers. and measured the inhibition zone distance measured from the paper disc to the outer inhibitory zone. In general, inhibition of bacterial growth refers to the general standard of antibiotics for E. coli, namely ampicillin with a range <11 mm insensitive, 12-13 mm sensitive enough and> 13 mm very sensitive [4].

3. Result and Discussion

Phytochemical assay of Green Macro Algae. Phytochemicals compounds of green macro algae sample test . is presented in Table 1 as follows

Table 1. Phytochemicals compounds of Green Macro Algae

Compound of Phenolic	U. reticulata	U. lactuca	C. racemosa
steroid	-	+++	-
terpenoid	++	-	+++

saponin	-	+	-
phenolic	+++	++	+++
alkaloid	+	+	+
flavonoid	-	-	+++

Note :
- \- : no sediment / color
- \+ : provide sufficient sediment/color
- ++ : provides moderate sediment / color
- +++ : provides lots of sediment / color

The results of the analysis of phytochemical compounds in macro algae U. reticulata consisted of terpenoids, phenolic, and alkaloids, while U. lactuca consisted of steroids, saponins, phenolics, and alkaloids. In C.racemosa consists of terpenoid, phenolic, alkaloid and flavonoid compounds. Differences in the content of phytochemical compounds of several types of seaweed may be caused by several things such as differences in type/species, place of life, season, and several other factors. This follows that there are various factors that can affect the content of phytochemical compounds, including species, varieties, growing conditions, seasonal variations, processing, and storage methods [5]. Environmental factors can also affect the content of phytochemical compounds such as cultivation location, altitude, temperature, sun exposure, rainfall, climate, and soil. These factors can affect secondary metabolites qualitatively and quantitatively, so that their bioactivity can vary [6]. Phytochemical test is a method of testing the content of secondary metabolites in seaweed [7].

The phytochemical test is a test to determine the types of secondary metabolite compounds or groups of compounds contained in the extract. Identifying the content of secondary metabolites is an essential first step in research to find new bioactive compounds from natural substances that can be precursors for the synthesis of new drugs or prototypes for certain active drugs. specifics for the qualitative test [8]

Antibacterial activity of Green Macro Algae. Green macroalgae sample were tested for the ability of its inhibitory activity against *E. coli* obtained the results presented in Figure 1 as follows.

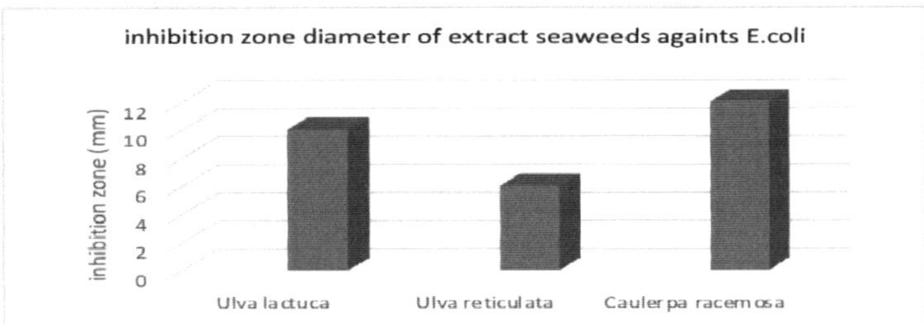

Figure 1. Inhibitory diameter zone of seaweed extract to E.coli bacteria
Note: *range* < 11mm insensitive, 12-13 mm sensitive enough, and > 13 mm very sensitive (4).

Antibacterial activity test of several types of green macroalgae in inhibiting bacteria E.coli growth resulted in C.racemosa seaweed having an inhibition zone diameter of 12.0 mm including quite sensitive category, then U.reticulata. has an inhibition zone diameter of 6.0 mm including the insensitive category, and U. lactuca has an inhibition zone diameter of 10 mm. The presence of phenolic compounds and their derivatives (flavonoids) contained in each type of seaweed maybe potential for inhibition E.coli bacteria growth, whereas the phytochemical compounds contained in

C.racemosa consist of phenolic compounds, flavonoids, alkaloids, and tannins. These flavonoid compounds can inhibit of pathogenic bacteria causing poisoningand diseases in humans, such as Escherichia bacteria coli [9].

Phenolic compounds and their products (flavonoids) are work by disrupting the function of the cytoplasmic membrane. The presence of these phenolic compounds can cause damage to the cytoplasm. H ions from phenol compounds and their derivatives will attack polar groups (phosphate groups). Phospholipid molecules in the bacterial cell wall will break down into glycerol, carboxylic acid, and phosphoric acid. Phospholipids cannot maintain the shape of the cytoplasmic membrane. As a result, the cytoplasmic membrane will leak, and the bacteria will experience growth inhibition and even death. Flavonoids prevent the formation of energy in the cytoplasmic membrane and inhibit bacterial motility, which also plays a role in antimicrobial action

The alkaloid compounds, flavonoids, steroids/triterpenoids, and tannins found in wild seaweed extracts are thought to be active as antifungal and antibacterial compounds [10]. The main chemical compounds with antibacterial properties are phenols, alcohols, halogens, heavy metals, dyestuffs, detergents, quaternary compounds, acids, and bases. Phenol compounds can interact with the components of the bacterial cell wall to cause permeability in bacterial cells and can also diffuse into cells so that bacterial growth is inhibited or dead; this compound can also penetrate the membrane and interact with genetic material so that bacteria undergo mutation [11]. The biological activity of alkaloid compounds is caused by alkaline groups containing nitrogen. When in contact with bacteria, the existence of this base group will react with amino acid compounds that make up the bacterial cell wall and bacterial DNA, which is the main constituent of the cell nucleus, which is the center of regulation of all cell activities. This reaction occurs because chemically, an alkaline compound will act with an acid compound; in this case, it is an amino because most amino acids have performed with alkaline groups of alkaline compounds. Changes in the composition of these amino acids will change the genetic balance in DNA acids so that bacterial DNA will be damaged. Damage to DNA in the nucleus of bacterial cells will encourage lysis in the cell nucleus so that cell damage will occur. Cell damage causes bacterial cells to be unable to metabolize to experience lysis (destroyed) [12].

The presence of phenolic compounds with hydroxyl groups attached to aromatic rings is a compound that is effective as an antioxidant and antibacterial because these compounds can reduce free radicals by giving hydrogen atoms (proton donors) from the hydroxyl groups to free radicals [13]. green algae (Ulva sp.)have activity in inhibiting growth of Escherichia coli bacteria and. This is becausepresence of secondary metabolites including triterpenoids, flavonoids, and saponins [14]

Conclusion. This research concludes that (1) Phytochemical compounds of *U.reticulata* are terpenoid, phenolic, and alkaloid. In *U. Lactuca* are steroid, saponin, phenolic, and alkaloid. Phytochemical compounds of *C. racemos*a are phenolic, flavonoid, alkaloid, and terpenoid. (2) *C. racemose* seaweed can inhibit *E.coli* bacteria with an inhibition zone diameter of 12.0 mm, including a susceptible category. Seaweed type U. lactuca has a 10 mm inhibition zone, and U.reticulata has a 6.0 mm inhibition zone diameter. including the sensitive category

Acknowledgment. Thank you to all those who have helped carry out this research Bacteriology and Mycology Laboratory of the Faculty of Veterinary Medicine, ,and Laboratory Genetic Resource Udayana University

References.

[1] Bachtiar, S.Y., W. Tjahjaningsih, and N. Sianita.2012.Effect of brown algae extract (Sargassum sp.) on Eschericia coli bacteria,"*J Mar Coast Sci*, vol. 1, no. 1, hal. 53–60, http://journal.unair.ac.id/download-fullpapers-53-60.pdf

[2] Eahamban,K and J. M. Antonisamy.2012 "Preliminary Phytochemical, UV-VIS, HPLC and Anti-bacterial Studies on Gracilaria corticata J. Ag," *Asian Pac. J. Trop. Biomed.*, vol. 2, no. 2

SUPPL., doi: 10.1016/S2221-1691(12)60275-5.

[3] Choironi,N.A and M. S. Fareza.2018. "Phytochemical Screening and Antibacterial Activity of Ethanolic Extract of Syzygium samarangense Leaves," *J. Kartika Kim.*, vol. 1, no. 1, hal. 1–4, 2018, doi: 10.26874/jkk.v1i1.2.

[4] P. Rajasulochana, P. Krishnamoorthy, and R. Dhamotharan. 2013. "An investigation on the antioxidants, antifungal and antibacterial of the Kappaphycus alvarezii," *Res. J. Pharm. Biol. Chem. Sci.*, vol. 4, no. 1, hal. 586–594

[5] Ate, J.N.B ., J. F. da Costa, and T. P. Elingsetyo S.. 2017. "Analysis of Nutritional Content of Gracilaria Edule (S.G. Gmelin) P.C. Silva and Gracilaria Coronopifolia J. Agardh for Economic Development of Coastal Communities," J. Health Sciences.,., vol. 5, no. 2, 94–103, doi: 10.30650/jik.v5i2.57.

[6] Riwanti,P., A. Kusuma, and R. Andayani. 2021 "ANTI-OXIDANT ACTIVITY OF 96 % Sargassum polycystum DPPH (2 , 2-diphenyl-1-picrylhydrazyl) METHOD USING UV-VIS SPECTROPHOTOMETRY" vol. I, no. 2, hal. 33–39,

[7] Lantah, P.L., L. A. Montolalu, dan A. R. Reo.2017, "PHOTOCHEMICAL CONTENT AND ANTIOXIDANT ACTIVITY OF SEAWEED METHANOL EXTRACT Kappaphycus alvarezii," *Media Teknol. Has. Perikan.*, vol. 5, no. 3, hal. 73, 2017, doi: 10.35800/mthp.5.3.2017.16785.

[8] Saleh, A.A.G.A., A. Asnani, and S. Suwarjoyowirayatno.2019 ""TEST OF Phytochemicals and ANTIOXIDANT ACTIVITY OF CENDOL FORMULATED FROM SEAWEED (Kappaphycus alvarezii) AND SAGO FLOUR (Metroxylon sagus Rottb.). *J. Fish Protech*, vol. 2, no. 1, hal. 38,doi: 10.33772/jfp.v2i1.6467.

[9] Sukmawati,S. 2021. Antibacterial Activities of Seaweed (Eucheuma cottonii) Extract Against Escherichia coli Bacteria," *Berk. Sainstek*, vol. 9, no. 1, hal. 6, 2021, doi: 10.19184/bst.v9i1.18184.

[10] Wiyanto, D,B,D,B. 2010."Antibacterial Activity Seaweed Extract Kappaphycus alvarezii and Eucheuma denticullatum to Aeromonas hydrophila and Vibrio harveyii Bacteria," *J. Kelaut.*, vol. 3, no. 1, pp. 1–17

[11] Sidauruk, S.W., N. Ira Sari, A. Diharmi, and I. Arif. 2021."Antibacterial Activity of Sargassum plagyophyllum Extract against Bacteria Listeria monocytogenes and Pseudomonas aeruginosa," *J. Pengolah. Has. Perikan. Indones.*, vol. 24, no. 1, hal. 27–37, doi: 10.17844/jphpi.v24i1.33417.

[12] Cox, S., N. Abu-Ghannam, and S. Gupta. 2010. "An assessment of the antioxidant and antimicrobial activity of six species of edible Irish seaweeds," *Int. Food Res. J.*, vol. 17, no. 1, hal. 205–220

[13] Febrianto, W., A. Djunaedi, S. Suryono, G. W. Santosa, dan S. Sunaryo. 2019. "Potential Antioxidant of Seaweed Gracilaria verrucosa from Gunung Kidul Beach, Yogyakarta," *J. Kelaut. Trop.*, vol. 22, no. 1, hal. 81, doi: 10.14710/jkt.v22i1.4669.

[14] Liswandari, M.S 2018. Antibacterial Activity Test of Green Alga (Ulva sp.) From Sorido Biak Beach to Escherichia coli and Staphylococcus aureus Bacteria. 2018. *J. Farm. Medica/Pharmacy Med. J.*, vol. 1, no. 1, hal. 9–15, 2018, doi: 10.35799/pmj.1.1.19646.

"Hulun Hyang" Farmer Group Asset Mapping Using Asset-Based Community Development Methode as an Ex-situ Edelweiss Flower Conservation in the Bromo Tengger Semeru National Park Area (Case Study: Edelweiss Park, Wonokitri Village)

Melati Julia Rahma*, **Soemarno Soemarno, Jati Batoro**

Brawijaya University, Malang, Indonesia

*melatijuliar@student.ub.ac.id

Abstract. One of the endemic flora that has become an icon in the Bromo Tengger Semeru National Park (TNBTS) area is the Edelweiss Flower. The existence of tourism developments makes the demand for Edelweiss flowers as souvenirs even more excellent. It affects the economic value of these commodities. It affects the population decline and threatens its availability for the Tengger indigenous people who use Edelweiss flowers as offerings that other components cannot replace. Since 2017, the Wonokitri Village community has collaborated with Bromo Tengger Semeru National Park (TNBTS) area to form the Hulun Hyang farmer group to conserve Edelweiss Flower ex-situ. The Hulun Hyang Farmer's Group has also obtained a breeding permit for Edelweiss Flowers from the Ministry of Environment and Forestry regarding wanawiyata widyakarya, which is realized through the development of the Edelweiss Park. It is essential to increase the capacity of managers in Edelweiss Park Development to ensure the sustainability of their activities. In this case, the Asset Based Community Development (ABCD) approach is considered suitable to be applied as a methodology to map assets owned by the Hulun Hyang Farmer Group. This asset mapping is hoped that it can encourage members of the Hulun Hyang Farmer Group to find out detailed information on their capital, which can later be used as a basis for formulating programs to improve skills and collective understanding so that they can develop a sustainable Edelweiss Park agro-ecosystem independently.

1. Introduction

The Bromo Tengger Semeru National Park area (TNBTS) has a native ecosystem, a habitat for a variety of flora and fauna with sub-montane, montane, and alpine ecosystem types [1]. One of the endemic flora that has become an icon in the TNBTS area is the Edelweiss Flower. Specific climatic conditions in the TNBTS area meet the growing criteria for Edelweiss Flowers (*Anaphalis spp*). The existence of tourism developments in the Bromo Tengger Semeru National Park (TNBTS) area makes the demand for Edelweiss flowers as souvenirs even more excellent. It affects the economic value of these commodities, and high market demand makes people interested in getting them. However, the sale of Edelweiss Flowers has been carried out illegally because the flower is one of the protected flora. The number that continues to decrease in their natural habitat can threaten the existence of Edelweiss Flowers [2]. Edelweiss flower also affects the traditional ceremonial activities carried out by the Tenggerese people because this flower is an essential component in offerings that cannot be replaced [3].

Residents of Wonokitri Village, Tosari District, Pasuruan Regency, formed a farmer group called Hulun Hyang (Servant Sang Hyang Widhi) in 2017. This farmer group was specially trained by the Bromo Tengger Semeru National Park Center (BBTNBTS) to encourage local communities to carry out conservation efforts by developing the Edelweiss Park as a means of ex-situ conservation and a display case for Edelweiss flower cultivation. The Hulun Hyang Farmers Group also officially obtained the Edelweiss (Anaphalis spp) flower breeding permit from the Ministry of Environment and

Forestry. Ka. BP2SDM:No.6361/MENLHK-BP2SDM/LUH/OTL.0/7/2019 regarding Wanawiyata Widyakarya. Wana Widyakarya itself is a business model in the forestry and or environmental sector owned and managed by community groups or individuals set by the Minister of Environment and Forestry as a pilot, training place, and internship for other communities. Wanawiyata Widyakarya is intended to provide quality learning facilities for the community in the forestry/environmental business sector and give appreciation to community groups and individuals who have succeeded in developing businesses in the forestry/environment sector [4].

Based on the analysis of the situation presented, what needs to be addressed in the advanced stages of the development of Edelweiss park is the mapping of assets owned to strengthen the group foundation as further consideration for developing human resource capacity in managing sustainable agroecosystems. This development needs to be fostered endogenously through the active participation of every member of the Hulun Hyang Farmer Group to form a firmly rooted mechanism and responsive to existing local dynamics. In this case, the Asset Based Community Development (ABCD) approach is considered appropriate to be applied as a methodology to determine the capacity assets of the Hulun Hyang Farmer Group. This study aims to analyze and map the assets that exist in the Hulun Hyang Farmer Group, which can be carried out for the sustainable development of Edelweiss Park and can be carried out further research in making an empowerment strategy for the Hulun Hyang Farmer group a group of people who are economically and socially independent so that they can develop a sustainable Edelweiss Park agroecosystem.

2. Material and Methods

The research was conducted from October to December 2021 in Edelweiss Park, Wonokitri Village, Tosari District, Pasuruan Regency. The area of Wonokitri Village is 1,120,295 Hectares consisting of 887,598 Hectares of agricultural fields, 14 ha of shrubs, 212,556 hectares of Forest Protection and Nature Conservation, and 20,127 hectares of plantations. Details of the research location are shown in Figure 1.

Figure 1. Research Location

2.1. Material

Data mining was carried out by semi-structured interviews and FGDs, which required tools in the form of draft questions as a guide in the implementation of data mining, stationery, and cameras to document data in the field and voice recorders to assist researchers in carrying out data mining. The types of data used are primary and secondary data. Primary data was obtained during data mining in the field and carried out directly by researchers, while secondary data was obtained from farmer group documents or literature studies. Data mining was carried out on 30 members of the Hulun Hyang farmer group who are over 17 years old.

2.2. Methods

This study uses the Assets Based Community Development approach, a concept of community development based on local assets in an area. The area has assets that can then be developed to be used to solve problems in the region [5]. Qualitative data analysis was done by the triangulation method. The triangulation technique used in this research is source triangulation, method triangulation, and time triangulation [6].

IOP Conf. Series: Earth and Environmental Science 1131 (2023) 012013 doi:10.1088/1755-1315/1131/1/012013

3. Result and Discussion

Assets or capital that become resources for the community, there are at least six assets, namely physical capital, financial capital, environmental capital, technological capital, human capital, social capital, and other supporting assets in the form of spiritual assets. These assets are potentials that need to be developed to support society's welfare [7]. The assets owned by the Hulun Hyang Farmer Group are detailed in Table 1.

Table 1. Hulun Hyang Farmers Group Assets

Assets Type	Owned Capital	Information
Physic	Village Land	7000 m^2 for edelweiss garden construction
	Nursery House	2 units for edelweiss flower nursery up to 3 months
	Counter	1 unit for the purchase of entry vouchers and storage of goods belonging to the Farmer Group
	Gazebo	5 units as facilities for Edelweiss Park tourists
	Irrigation	1 unit as irrigation channel to support Edelweiss Flower conservation activities
	Lighting	1 unit as a lighting channel to support Edelweiss Flower conservation and tourism activities
	Internet Access	1 unit as communication access and promotion to support Edelweiss Flower conservation activities and tourism
Financial	Grant Fund	From Bank Indonesia for the construction of facilities and infrastructure in Edelweiss Park
	Entrance Voucher	For Rp. 10,000.00 per person which can be exchanged for tea or coffee in the cafe section.
	Cafe menu	Food or drink options that have been provided with a price range of Rp. 5,000.00 to Rp. 20,000.00
	Souvenir	Souvenirs from the cultivation of edelweiss flowers which are managed by members of the Hulun Hyang Farmer's group with prices ranging from Rp. 20,000.00 to Rp. 50.000,00
Enviroment	Type of soil	Typic hapludans with good drainage conditions, rough soil texture, slightly acidic pH, medium Cation Exchange Capacity with low Base Saturation. Landform Lungur Vulkan with Basalt Andesite Material. [8]
	Air temperature	15-25°C [9]
	Altitude of place	1600-3.600 masl [9]
	Rainfall	3.712 mm per year [9]
	Land slope	Mountainous relief with land slope >40%. [8]
	Flower Type	*Anaphalis javanica, Anaphalis longifolia, Anaphalis viscida* [10]
Technology	Preparation of planting media	For breeding, it is necessary to prepare the soil with a mixture of chicken manure in a ratio of 1: 1
	Seed preparation	Edelweiss flowers that have flowered over time will dry up and have blackish seeds. The seeds are then sown.
	Nursery	The seeding is done in a greenhouse to maintain the humidity of the air and water. Watering is carried out twice a day (morning and evening, does not get flooded). When it has grown into a sprout, then weaning is done. Where each sprout is carefully

Assets Type	Owned Capital	Information
		separated and transferred in polybags.
	Cultivation	After the Edelweiss Flower is three months old, then it is moved to the prepared land. The land is fertilized with a composition of 0,5 kg of chicken manure per plant.
	Plant Management	Fertilization with chicken manure once a month to maintain nutrients for plants. When the edelweiss flowers are moved to the field, they are only watered once a day in the dry season and during the rainy season they only rely on rainwater
	Harvest	The harvest period = one year, from May to August. Harvesting by trimming the flower stems and drying them for three days in a dry room. Puddles of water will cause the rotting of flowers due to fungus. After the flowers dry, the flower seeds are taken for the following seed material. The dried Edelweiss flowers are then assembled into various forms of souvenirs.
Human Resource	Membership	30 members
	Age	>30 years = 6 members ; < 30 years = 24 members
	Level of education	6 members = Bachelor ; 24 members = Senior High School
Social	Traditional ceremonies	Kasada, Karo, Leliwet, Entas-Entas and Unan-Unan
	Customary law	Edelweiss flowers (tana layu) are essential in several traditional ceremonies and cannot be replaced with other flowers.
	Country law	Law of the Republic of Indonesia Number 5 of 1990 concerning the Conservation of Biological Natural Resources and Ecosystems in Article 33. For those who violate this article, further details on fines and imprisonment are explained in Article 40.
Spiritual	Ancestral and religious values	There is a close relationship between environmental humans and their God in interpreting gratitude for the abundance of existing natural resources, which are maintained very well from generation to generation.

Natural resource assets are related to the ownership or control of natural resources such as land, plants, and water as production assets [11]. Given its nature that is difficult to change, farmers use natural resources according to their ownership. One of the essential natural resource assets for farmer groups is the environment's carrying capacity, which is very specific and supports Edelweiss Flowers to grow. Biological wealth in forests as part of natural resources plays a significant role in people's lives, especially in applying cultural values [12] [13]. In addition, social assets have an essential role in maintaining natural resources to preserve livelihoods [14]. The management of existing natural resources needs to be supported by the existence of institutions. Institutions are a collection of values, norms, and rules in a group of people that are used to achieve specific goals [15]. In the Hulun Hyang Farmer group, the goals that will be shared are related to the conservation of Edelweiss Flowers outside the Area so that the flowers remain sustainable because they are also related to the sustainability of the culture and customs of the Tengger people.

Physical assets are the next priority because farmer groups are aware that edelweiss flower conservation activities can also be linked to tourism activities. Physical assets are the basic infrastructure and goods needed to support livelihoods. In addition to infrastructure, physical assets also include ownership of equipment and infrastructure facilities for production activities. In a regional context, it can be in the form of road infrastructure, irrigation, and access to information and communication [11]. The infrastructure in question is public goods that everyone can access at no cost.

Figure 2 . Mapping of Hulun Hyang Farmer Group Assets

Financial assets are significant because they are the most liquid among other assets, so that they can be directly used or exchanged with other assets. Financial assets can be used to buy products or consumer goods. Social assets depend on the participation of individuals and group members in social networks through communication, knowledge, and support provided [16]. Communication, knowledge, and support for managing Edelweiss Gardens are essential because it determines farmer group members' perceptions of the ex-situ conservation area for Edelweiss Flowers. The perceptions formed will affect the forms of interaction that positively or negatively impact Edelweiss Park [17]; [18]. The subsequent domination of assets is human resource assets. Human resources are active agents to collect assets, exploit natural resources, build social, economic, and political relations, and bring about changes in the future [19].

Human resource development is significant because the adoption of technology and knowledge in community is very dependent on the capacity of the management. Several factors that influence the adoption of technology in the management of Edelweiss park are human resources related to age and education. The age of farmer group members dramatically influences their behavior towards technology adoption. Younger farmers will be more innovative and more open to new technologies. Age is an individual characteristic that plays a significant role in determining workability and work productivity [20]; [21]. Likewise, with the level of education, farmers with higher education tend to be more open to adopting new technologies [22]. According to [23], the essence of social capital lies in the ability of the community to work together to build a network to achieve common goals. This collaboration is characterized by reciprocal and mutually beneficial interrelationships and is built on trust, supported by positive and robust social norms and values. Where social norms and values are built based on values from ancestors that have been taught from generation to generation

4. Conclusion

The mapping of livelihood assets determines the forms of intervention that can be carried out to increase the role of Edelweiss Park as an Ex-Situ Edelweiss Flower Conservation area in the livelihoods of the Hulun Hyang farmer group. Asset mapping is visualized in a pie chart to see if each asset owned is related to the other. For the Hulun Hyang farmer group, the priority of asset use is environmental, social, spiritual, human, and financial assets.

5. References

[1] Purnomo, Traditional Environmental Conservation Practices in Java. Malang: UB Press, 2015.
[2] N. Maulidah, "Analysis of Maslahah and Mafsadah on the Sale and Purchase of Edelweiss Flowers (Anaphalis Javanica) at Mount Bromo Tourism (Bromo Tengger Semeru National Park)," Sunan Ampel UIN Surabaya, Surabaya, 2015.
[3] R. W. Hefner, Hindu Javanese: Tengger tradition and Islam. Princeton New Jersey: Princeton University Press, 1985.
[4] KEMEN-LHK, "Regulation of the Minister of Environment and Forestry of the Republic of Indonesia," 2016.
[5] G. P. Green and A. Haines, Asset Building and Community Development. United States.
[6] D. C. Junissa, "Construction of Agricultural Meaning And Reality For Milenial Entrepreneurs In Bandung Barat District," Communication Of Agricultural And Rural Development Postgraduate School Institute Of Agriculture Bogor, Bogor, 2021.
[7] Muhtar, "Community Development using local assets," Sosioconceptual, vol. 17, no. 01, 2012.
[8] Research and Development Agency of the Ministry of Agriculture, "Semi Detail Soil Map of Pasuruan Regency (Sheet 1608-12)," Ministry of Agriculture, Bogor, 2016.
[9] R. Hariyatmi and S. Ahmad, "Diurnal Butterfly Diversity (Sub Order: Rhopalocera) in Mount Bromo Complex, in: KPH Surakarta Karanganyar Regency in 2013," presented at the XI National Seminar on Biology Education FKIP UNS, Semarang , 2013.
[10] C. G. G. J. Van Steenis, The Mountain Flora of Java. Bogor: Center for Biological Research. Institute of Sciences (LIPI), 1978.
[11] I. Scoones, "Sustainable Rural Livelihoods: A Framework for Analysis," IDS Working paper Brightons, vol. 7, 1998.
[12] A. Nurhadi, B. Setiawan, and Baiquni, "Environmental Wisdom in Planning and Management of the Wonosari Forest, Ngawen District, Gunungkidul Regency," Journal of Humans and the Environment, vol. 19, no. 3, pp. 226–237, 2012.
[13] S. T. Salosa, S. A. Awang, P. Suryanto, and R. H. Purwanto, "Forests in the Life of the Hatam Community in the Arfak Mountains Nature Reserve," Journal of Humans and the Environment, vol. 21, no. 3, pp. 49–55, 2014.
[14] Z. Jele, "The Contribution of Small-Scale Timber Farming in Enhancing Sustainable Livelihood at Sokhulu.," University of South Africa, South Africa, 2012.
[15] S. A. Awang, Basic Principles of Institutional Analysis in Community Forestry Enterprises, in the Nunga Rampai Series on Community Forests: Farmers, Economics and Conservation. Aspects of Research and Ideas. Yogyakarta.: Debut Press, 2005.
[16] F. Ellis, Rural Livelihoods and Diversity in Developing Countries. Oxford: Oxford University Press, 2000.
[17] A. Kadir, S. A. Awang, R. H. Purwanto, and E. Poedjirahajoe, "Stakeholder Analysis of Bantimurung Bulusaraung National Park Management, South Sulawesi Province.," Journal of Humans and the Environment, vol. 20, no. 1, pp. 11–21, 2013.
[18] S. A. Raharjo, S. A. Awang, A. Pramusinto, and R. H. Purwanto, "The History of State Domination in Sandalwood Management in East Nusa Tenggara," Journal of Humans and the Environment, vol. 20, no. 1, pp. 1–10, 2013.
[19] D. A. Olaniyan and T. Okemaninde, "Human Capital Theory: Implication for Educational Development," European Journal of Scientific Research, vol. 24, no. 2, pp. 157–162, 2008.
[20] T. H. Handoko, Personnel Management and Human Resources. Yogyakarta: BPFE, 2001.
[21] A. Kadir, S. A. Awang, R. H. Purwanto, and E. Poedjirahajoe, "Analysis of Socio-Economic Conditions of Communities Around Bantimurung Bulusaraung National Park, South Sulawesi Province," Journal of Humans and the Environment, vol. 19, no. 1, pp. 1–11, 2012.
[22] S. W. Diggenaro, "Evaluation of the Livelihood Impact of a Micro-Irrigation Project in Zambia," The Ohio State University, Ohio, 2010.

[23] J. Hasbullah, Social Capital: Towards Excellence in Indonesian Human Culture. Jakarta: MRUnited Press, 2006.

Collaborative Management to support sustainable Community Forest

Hertasning Yatim[1*], Yusran Yusran[2], Supratman[2], Nur Zaman[3]

[1]Tompotika Luwuk University, Luwuk, Indonesia
[2]Universitas Hasanuddin, Makassar, Indonesia
[3]Lecturer of Agriculture Faculty of Technology Sulawesi University, Makassar.

*hertasningyatim70.hy@gmail.com

Abstract. The current state of social forestry, including community forestry (HKm) in Indonesia, is practically stagnant. Many factors cause stagnation in HKm management, including the main one being the interaction with the Forest Management Unit (FMU) as forest manager at the site level. This study aims to build an institutional model for sustainable forest management through analysis of the arena of action, community attributes, and forest management rules at the Nambo Community Forest, Central Sulawesi. The results of this research found that the collaborative management model between the government and the community is suitable to be applied in Community Forest management, because it provides greater opportunities for recognizing the rights of local communities to forest resources, as well as increasing the empowerment of rural communities.

1. Introduction

Community Forestry (CF) is one of the Ministry of Environment and Forestry's policies to reduce Indonesia's deforestation by involving the community and Village Forests and Community Forests Plantation. More than 50 million poor Indonesians live in and around forest areas who depend on forest resources.. The broad definition of Community Forestry is "initiatives, science, policies, institutions and processes intended to enhance the role of local communities in regulating and managing forest resources [8]. Community forestry, as a local institution and community-based natural resource management, has been studied in recent decades as a means of rural livelihoods [1]. For this study, community forestry is defined as an entity with an explicit mandate and legal authority to manage forest areas for the benefit of local communities [12]. Forest is a form of shared resource that is important for rural communities [7].

Many developing countries have adopted co-management in anticipation of the problem of forest degradation [2]. Co-management methods are designed to empower and secure the livelihoods of local communities. it is necessary to have a common perception of institutional models that are in accordance with local social, economic and cultural characteristics. The development of appropriate forest management institutions will lead to good forest management performance and can be designed through the development of appropriate incentives [6].

2. Methodology

This paper used a qualitative descriptive research design with a case study approach. This research was conducted in the Production FMU Model Toili Baturube area, Banggai Regency, Central Sulawesi. The data collection and processing process is carried out in the FMU area and the Koyoan Community Forest (CF) area, as well as the offices of related agencies in Banggai Regency and the Ministry of Environment and Forestry, namely the Production FMU Office Model Toili Baturube, Banggai Regency and the Environment and Forestry Service of Central Sulawesi Province. Data and information collected in this study include primary data and secondary data. Preliminary data were obtained through a structured interview approach using a questionnaire with respondents, in-depth interviews with community leaders, agencies/agencies, and related Community Social Institutions (NGOs), and observations of objects in the field related to the research data topic. Sources of data are informants and stakeholders directly involved in the management of Production FMU Toili Baturube and the Community Forest Management Forest Farmers Group. It spread over four villages, namely Nambo Lempek Village, Nambo Padang Village, Koyoan Village, and Lontio Village. Each village will be taken as many as 30 farmer respondents, so 120 respondents. The selection of respondents was made by purposive sampling. Secondary data is obtained through literature studies on several documents, such as maps, planning documents, activity reports, archives, or other types of documents concerning Forest Management Units (FMU) and Community Forests (CF).

Data analysis in the institutional development framework generally refers to the study of institutional development (IAD) proposed by Ostrom. By describing the rules used, understanding the arena of action and patterns of interaction; analyze the content of the regulations concerning the structure of the action situations formed and the resulting performance. In this arena, actors (e.g., forest managers, owners) come together to determine forest management actions and what will be taken to address specific situations [5]. In this study, the written documents analyzed were the laws and regulations directly or indirectly related to the CF regulation. This study adopts a line of thought and an analytical approach, as described in the IAD framework (see Figure 1). According to Ostrom [17], the IAD framework can help institutional analysts identify the key variables that shape the structure of participants' action situations.

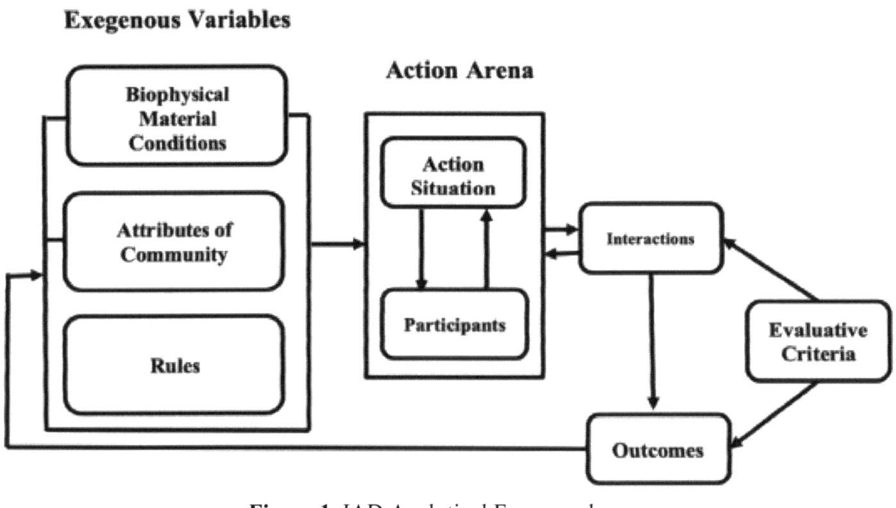

Figure 1. IAD Analytical Framework

Qualitative Descriptive through the use of Ostrom's Design Principles and Stakeholder Analysis to Formulating an institutional collaboration model in HKm management that supports the sustainability of sustainable forest management. Ostrom designed 8 (eight) principles that characterize the configuration of rules used by CPR institutions that are sustainable and involve collective action [17]. As for the framework of this research, it can be seen in Figure 2.

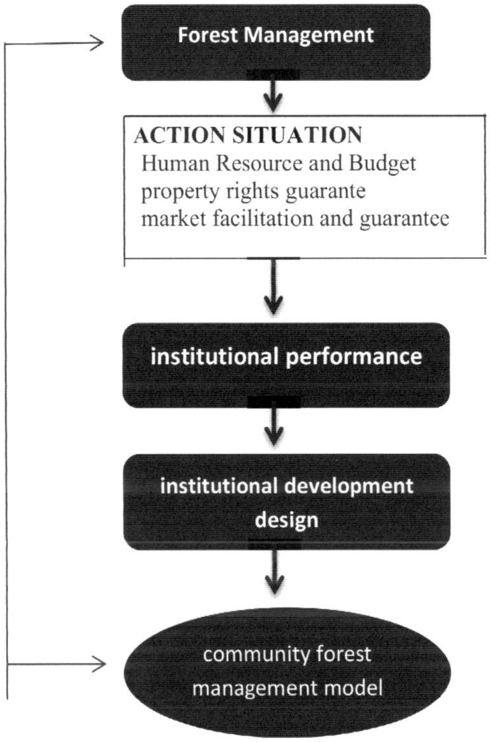

Figure 2. Research Framework

3. Result and Discussion

As shown in Figure 1, the IAD framework represents an institutional process, in which different variables or procedures exist. At the heart of the IAD framework is the situation of action, an abstraction from the decision-making environment in which individuals and actors interact to make choices that determine the outcome of several aspects of the policy question. Individual decisions and collective outcomes are influenced by relevant individuals' beliefs and incentives shaped by social responsibilities and expectations [6]. According to Ostrom [17], the action arena consists of situations of activities and participants. In this case, participants are composed of three variables: the preference variable for the structure of the action situation, how to process and use information, and the selection criteria used in making decisions. Based on the types of variables, the structure of the action situation for forest use is based on Permenhut No. P. 37/2007 jo. P. 18/2009 jo. P.88/2014 concerning Community Forestry, as shown in Table 1.

Table 1 The situation of action on community forest utilization schemes in *Production FMU* Toili Baturube

Participants	Form of Authority	
	PP 52/2011 jo. PP 88/2014	**UU 23/2014**
Ministry of Forestry	Assisting facilitation and mentoring	The implementation of government affairs in the forestry sector is divided between the Central and the Provinces
Governor	Assisting facilitation and mentoring	Assisting facilitation and mentoring
Forestry Service	Assisting facilitation and operational	Assisting facilitation and mentoring
FMU	Site-level forest management	Site-level forest management
Community Forest Farmer Groups	Permit holder and entitled to get permission and assistance	The status remains the same
Ministry of Forestry	Financing for the implementation of CF	Financing for the implementation of CF
Governor	Financing for the implementation of CF	Financing for the implementation of CF
Provincial Forestry Service	Financing for the implementation of CF	Financing for the implementation of CF
Regent	Funding can be sourced by regional revenue and expenditure budget	Limited authority
District Forestry Office	Funding can be sourced by regional revenue and expenditure budget	Authority is withdrawn to the province
FMU	Obtain budget allocation funds from APBN and APBD	Obtain budget allocation funds from APBN and APBD
Community Forest Farmer Groups	Receive facilitation and assistance funds	Receive facilitation and assistance funds

The Central Sulawesi Provincial Government, as an extension of the Central Government in the regions, until now has not been well equipped with regards to regulations, budgets, and institutions. Institutions and organizations are two important components [15], [18]. Institutions are built to inhibit the emergence of mutually harmful behavior so that the goal of creating a prosperous society can be achieved [11]. Support in the form of a Governor Regulation concerning the implementation of social forestry facilitation is needed to provide social forestry facilitation performance guidelines. It aims to support the acceleration of facilitation for the community in preparing and developing social forestry enterprises, resolving tenure problems through social forestry, and arranging coordination, integration, and synchronization to improve parties' participation in supporting social forestry. It can be seen in figure 3.

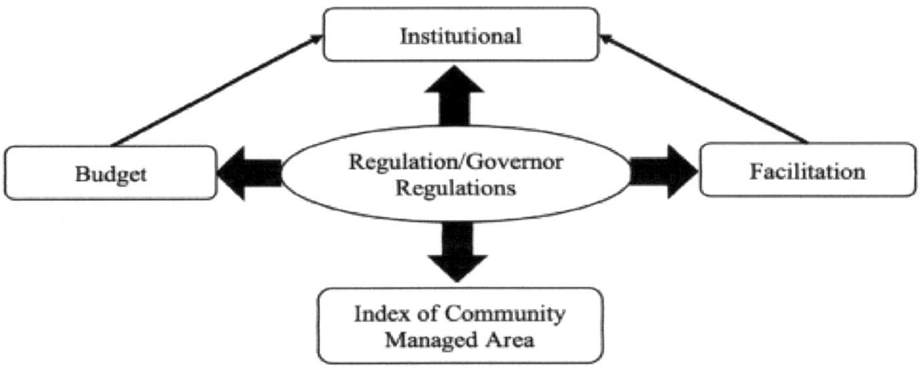

Figure 3. Linkage of Regulation, Budget, Facilitation, and Institutions

Collaborative management or co-management is defined as the agreement of two or more stakeholders to share information, roles, functions, and responsibilities in a mutually agreed relationship and partnership mechanism [4]. Co-management is also defined as the sharing of power and responsibility between the government and local users of resources [3]. HKm is a practice towards collaboration, where the most important community components are community rights and local knowledge and wisdom, are accommodated in forest management and community internalization occurs as forest managers and guards.[13] explained the principles in building collaborative forest resource management, namely (1) a combination of technical and social approaches is the key to the success of forest area management, (2) the strength of mutual trust and respect between parties, (3) understanding (mutual understanding) and agreement on the goals to be achieved together, (4) division of roles and responsibilities of each party, (5) fulfillment of the interests of each party, and (6) frequency of communication.

The essence of good forestry governance is how the actors involved can organize their inter-relationship mechanisms through various instruments controlling these inter-relational mechanisms to achieve common goals in the management and management of forest resources and the provision of sustainable forestry goods and services. There are 5 influential actors in the realization of good forestry governance [19], namely: 1) politicians and other pressure groups, 2) policy makers, 3) service provider organizations, 4) frontline professionals, and 5) clients. Based on the ideal institutional criteria and the roles of influential actors in the realization of good forestry governance as stated by the World Bank [19], as well as the collaborative management principles of forest resource management mentioned [13], it can be formulated collaborative institutions for HKm management as can be seen in seen in figure 4. The institutional collaboration model for the management of the Koyoan HKm in the KPHP area The Toili Baturube model basically integrates central government policies with an ideal institutional process that involves a number of actors and their inter-relationship mechanisms, and the principles of collaboration in forest resource management .

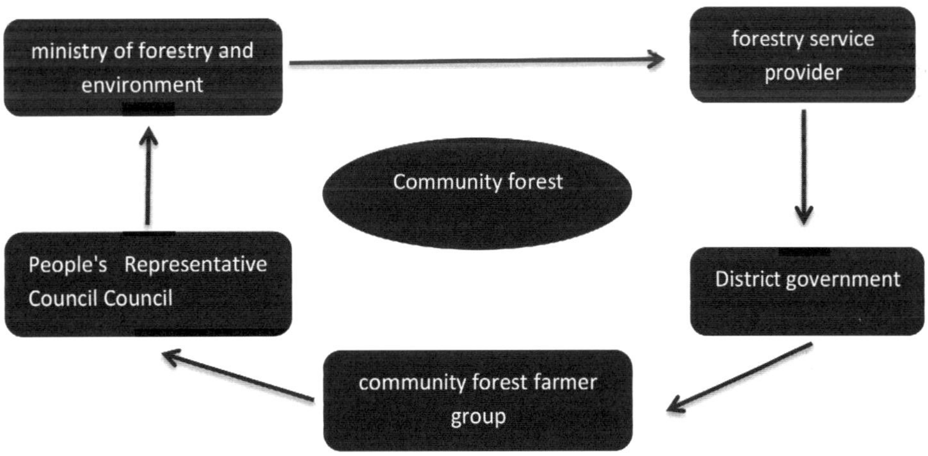

Figure 4. Community forest collaborative management model

4. Conclusion

This article analyzes how actors (in this case forest farmer groups and forest management units as well as the central government) are involved in interaction patterns in managing community forestry, as well as how regulation as a bureaucratic tie-in has implications for wider forest planting schemes at the local level. This study aims to build an institutional model for sustainable forest management through analysis of the arena of action, community attributes, and forest management rules at the Nambo Community Forest, Central Sulawesi. To achieve sustainable forest management, the recognition and incorporation of local institutions in forest policy formulation is very important because it has great potential for collective action and has the characteristics of common resources required for sustainable management.

The results of this study found that the institutional model of collaborative-management between the government and the community is suitable to be applied in community forest management, because it provides greater opportunities for recognition of local community rights to forest resources, as well as improvements in village community empowerment. Collaboration indicators are indicators of increased trust, improved communication and relationships, and reduced conflict, producing benefits [14]. Government facilitation will further generate trust and increase relationships among stakeholders and resource managers, greater participation, accountability, and transparency among the parties [20].

References

[1] Adhikari, B., Williams, F., Lovett, J.C. 2007. Local benefits from community forests in the middle hills of Nepal. Forest Policy and Economics, 9(5), 464-478.

[2] Adhikari KP., Goldey P. 2010. Social capital and its —downside‖: The impact on sustainability of induced community-based organizations in Nepal. *World Development 38 (2): 184-194.*

[3] Berkes F, 2002. Cross-scale institutional linkages: perspectives from the bottom up. In: Ostrom E., Dietz T., Dolsak N., Stern PC., Stonich S., Weber EU. (Eds.), *The Drama of the Commons*. National Academy Press, Washington DC, pp. 293–321

[4] Borrini-Feyerabend G., Farvar MT., Nguinguiri JC., Ndangang V. 2000. Comanagement of natural resources: Organizing negotiation and learning by doing Kasparek, Heidelberg

[5] Brodrechtova, Y., Navrátil, R., Sedmák, R., Tuček, J. 2018. Using the politicized IAD framework to assess integrated forest management decision-making in Slovakia. Land Use Policy, 79(1), 1001-1013.

[6] Cole, D. H., Epstein, G., McGinnis, M. D. 2019. The utility of combining the IAD and SES frameworks. International Journal of the Commons, 13(1), 244-275.

[7] Das N. (2010). Incidence of forest income on reduction of inequality: evidence from forest dependent households in milieu of joint forest management. Ecological Economics 69: 1617–1625

[8] Friedman, R. S., Guerrero, A. M., McAllister, R. R. J., Rhodes, J. R., Santika, T., Budiharta, S., Wilson, K. A. 2020. Beyond the community in participatory forest management: A governance network perspective. Land Use Policy, 97(January 2019), 104738.

[9] Gunningham, N. 2009. The new collaborative environmental governance: the localization of regulation. Journal of Law and Society, 36(1), 145-166

[10] Guy PB. 2000. Institutional theory: Problems and prospect. Political Science Series (69*)*. Vienna: Institute for Advance Studies.

[11] Kasper, W., Streit, M. E. 1996. Institutional Economics: Social Order and Public Policy. Edwar Elgar. Cheltenham, UK. Northampton, MA, USA.

[12] Krogman, N. and T. Beckley. (2002). Corporate "Bail-Outs" and Local "Buyouts": Pathways to Community Forestry? Society and Natural Ressources 15:109–127

[13] Mattessich, P.W.; M. Murray-Close; and B.R. Monsey. 2001. Collaboration: What Makes it Work. Amherst H.Wilder Foundation. Saint Paul, Minnesota

[14] McIntyre, K. B., Schultz, C. A. 2020. Facilitating collaboration in forest management: Assessing the benefits of collaborative policy innovations. Land Use Policy, 96(

[15] North, Douglas C. 1990. Institutions, Institutional Change, and Economic Performance. Cambridge University Press. New York the USA.

[16] Ostrom E, Crawford S. 2005. A grammar of institutions. In Ostrom E (Ed.) Understanding Institutional Diversity. Princeton (US): Princeton University Press.

[17] Ostrom, E., 1990. Governing the Commons: The Evolution of Institutions for Collective Action. Cambridge University Press.

[18] Rodgers, Gerry. 1994. Workers, Institutions, and Economic Growth in Asia. International Institute for Labor Studies. Geneva. Switzerland

[19] World Bank. 2005. Forest Institutions in Transition: Experience and Lessons from Eastern Europe. PROVOR World Bank. Washington. United Stated.

[20] Wondolleck, J.M., Yaffee, S.L., 2000. Making collaboration work: Lessons from innovation in natural resource management. Island Press

IOP Conf. Series: Earth and Environmental Science 1131 (2023) 012015 doi:10.1088/1755-1315/1131/1/012015

Growth and fruit yield of sponge gourd (*Luffa acutangula*L.) as influenced by combination of plant spacing and chicken manure

A R Fahreza, N Aini*

Universitas Brawijaya, Malang, Indonesia

*nra-fp@ub.ac.id

Abstract. This study aims to determine the growth and yield of sponge gourd (*Luffa acutangula* L.) by adjusting the spacing and doses of chicken manure. A field experiment was conducted at experimental station of Brawijaya University, Malang Regency, East Java. This study used a Randomized Block Design (RBD) with one combination factor. The treatment combination is plant spacing + chicken manure doses, which consist of: P_0 = 60 x 50 cm + 4 t ha^{-1}, P_1 = 40 x 50 cm + 4 t ha^{-1}, P_2 = 60 x 50 cm + 6 t ha^{-1}, P_3 = 40 x 50 cm + 6 t ha^{-1}, P_4 = 60 x 50 cm + 8 t ha^{-1}, P_5 = 40 x 50 cm + 8 t ha^{-1}. Obsevation variables are number of Leaves, plant length, leaf area, fruit weight per fruit and fruit weight per plant, number of fruits, and yield of fruit. The results showed that the treatment with a spacing of 40 × 50 cm + 8 t ha^{-1} chicken manure significant increasing the growth and fruit yield of sponge gourd. However, the fruit weight per plant with a spacing of 60 × 50 cm + 8 t ha^{-1} chicken manure had a higher value than the treatment with a spacing of 40 × 50 cm + 8 t ha^{-1} chicken manure. R/C ratio of all treatments are more than one. The treatment with a spacing of 40 × 50 cm + 8 t ha^{-1} chicken manure is not only feasible to cultivate, but can also be used for sustainable agriculture.

1. Introduction

Sponge gourd (*Luffa acutangula* L.) is one of the annual fruit vegetables that grows vines. Sponge gourd contains nutrients such as vitamins, fiber and minerals. In addition, gambas also contain several chemical compounds that are good for health such as triterpene saponins, luffein, citrulline, and cucurbitacin. The changes in people's lifestyles that lead to healthier lifestyles cause vegetable consumption to increase. However, at this time the sponge gourd plant is rarely cultivated or used as a side crop, so that there is still not much production of sponge gourd in Indonesia. The Efforts that can be made in overcoming this problem are cultivating gambas plants with appropriate cultivation techniques. For this reason, this study will try to use spacing and application of chicken manure. Plant spacing is related to plant population per unit area, and competition between plants in the use of light, water, and nutrients [1].

Planting distances that are too narrow will give less than optimal results because of the competition between the plants themselves. Plant spacing arrangements greatly affect plant growth and yield during fruit formation. Technology other than plant spacing in sponge gourd cultivation is the application of manure. The application of manure can improve both soil aggregates and aeration so as to optimize the availability and balance of nutrients in the soil [2]. Chicken manure has advantages over other manures, namely it has a higher nitrogen content and is more easily absorbed by plants [3]. Therefore, this study is expected to determine the appropriate spacing and dosage of chicken manure, so that higher sponge gourd yields will be obtained. This study aims to determine the effect of spacing and dose of chicken manure on the growth and yield of sponge gourd.

2. Material and Methods

2.1 Material

The tools used in this study were hoe, stationery, camera, hose, analytical scale, LAM, seedling tray, stake (bamboo), caliper, gawar rope, raffia rope, gembor and scissors. The materials used were sponge gourd seeds of Prima F1 variety, chicken manure, NPK fertilizer, pesticides, water and planting media (a mixture of soil and chicken manure).

2.2 Methods

General. The study was conducted at the Agrotechnopark Universitas Brawijaya at an altitude of ± 330 meters above sea level with an average temperature of 27 to 29 °C and rainfall between 102 to 97 mmmonth^{-1}. The soil condition on the land is clayey loam. The research area has a size of 196.91 m^2 which is detailed in a length of 20.3 m and a width of 9.7 m. For the size of each plot has a width of 1.8 m and a length of 2.8 m. The plot of land with a spacing of 60 cm × 50 cm contained 20 plants and a plot of land with a spacing of 40 cm × 50 cm contained 28 plants.

Research Implementation: Nurseries include preparation of planting material, preparation of planting media, nursery seeds and move crops.Preparation of land for tillage, making plots in the form of mounds, giving chicken manure, spacing and planting holes, Planting, will be carried out in the afternoon in order to avoid death in plants due to the influence of high temperature, fertilization include chicken manure and NPK fertilizer. The maintenance carried out includes replanting, watering, weeding, installing stakes, binding of tendrils, pruning and controlling plant pests and last is harvest.

Growth observations started at the age of 1 WAP (Week After Planting) and observations were made every week until 5 WAP, with an observation interval of 1 week. The number of samples used in each growth observation was 3 samples for each treatment. Number of Leaves: The number of leaves counted is the leaves that are already green and are actively photosynthesizing.Plant Length: Measures plant length using a meter scale from ground level to the highest plant canopy. Leaf area: leaf area per plant was calculated using a leaf area meter (LAM) using 30 leaves on the periphery of plants mixed (large, medium and small), then measured by LAM and the result was divided by 30 to find the average, then the result is multiplied by the number of leaves per plant. Leaf area measurements were carried out when heading to the generative phase or at the age of 28 DAP.Fruit weight per fruit and fruit weight per plant: Observation of fresh fruit weight per fruit and fruit weight per sample plant was carried out six times by weighing fresh fruit from all sample plants. Number of Fruits: Observation of the number of fruits per sample plant was carried out for six times by counting the number of fruits from all sample plants with 6 harvests. Yield: The weight of the fruit is weighed after harvesting the entire harvest plot and weighs the yield of the harvest. The yield of each harvest was calculated in order to obtain the yield with a harvest plot area of 1.2 m × 1 m and yield per ha.

Statistic. The results obtained were analyzed using Microsoft Excel software for further processing. Furthermore, the observational data was analyzed for variance to test the effect of the treatment, with the F test at the 5% level. If the treatment has a significant effect, then it is continued with the Least Significant Difference (LSD) further test at the 5% level.

3. Results and Discussion

Plant growth. The results of this study showed that the combination treatment between spacing and dose of chicken manure gave a significant effect on plant length at all ages, except 7 days after planting. At 28 days after planting, it was shown that the plant spacing of40x50 cm + 8 tha^{-1}chicken manure was significantly different in increasing plant length compared to other treatments (Tabel 1). The average length of the longest plant in the plant spacing of40x50 cm + 8 t ha^{-1}chicken manure treatment was 278.3 cm. The shortest average plant length in the treatment of 40x50 cm spacing + 4 t ha^{-1} of chicken manure was 202.0 cm. This is because a smaller spacing will increase the length of the plant due to competition in receiving sunlight. In addition, at a narrow planting distance, the space required is also too narrow so that the plants grow elongated upwards due to etiolation. This is in accordance with the opinion of [4], the increase in plant length at a smaller spacing is due to the closer plant canopy, resulting in a decrease in the intensity of light received. The closer the spacing, the faster the growth of plant length because the plants try to find more sunlight.

The results showed that the combination treatment between spacing and doses of chicken manure had a significant effect on the number of leaves at all ages, except 7 days after planting. At 28 days after planting, it was shown that the plant spacing of40x50 cm + 8 t ha^{-1}chicken manure gave the highest average number of leaves, which was 27.0 sheets compared to other treatments (Table 1). The least number of leaves in the treatment of 40x50 cm spacing + 4 t ha^{-1} of chicken manure is 20.5 sheets.The number of leaves is strongly influenced by chicken manure, because even though the spacing is smaller and the application of the right chicken manure will encourage leaf growth due to the content of nutrients, especially sufficient nitrogen. This is in accordance with [5], a plant will grow and develop properly if the nutrients needed are present and available enough to be absorbed by plant roots.

This study showed that the combination treatment between spacing and dose of chicken manure had a significant effect on leaf area per leaf and leaf area per plant. The highest mean of leaf area per plant at 40x50 cm spacing + 8 t ha^{-1} of chicken manure was 8,414.6 cm^2. The lowest mean of leaf area per plant in the treatment of 40x50 cm spacing + 4 t ha^{-1} of chicken manure was 3,384.7 cm^2. [6] stated that the large leaf surface increases the capture of light and CO_2 more effectively, so that the rate of photosynthesis increases, then the photosynthetic results are translocated to areas of vagetative utilization, namely roots, stems, leaves, which affect plant growth and development. According to [7], the value of leaf area depends on the method of cultivations, for example the use of plant spacing, the use of close spacing with the application of nitrogen contained in high chicken manure will increase the value of leavesweight.

Table 1. Average Plant Length, Number of Leaves and Leaf Area of sponge gourd at 28 Day After Planting

PlantSpacing (cm)+ Chicken Manure (t ha^{-1})	Plant Length (cm)	Leaf Number (sheetsplant^{-1})	Leaf Area (cm^2 plant^{-1})
(60x50) + 4	202.0 a	20.5 a	3384.7 a
(40x50) + 4	209.1 a	22.7 ab	5339.2 b
(60x50) + 6	203.5 a	22.3 ab	5306.0 b
(40x50) + 6	223.4 a	23.5 b	5553.4 b
(60x50) + 8	219.6 a	23.6 b	5867.1 b
(40x50) + 8	278.3 b	27.0 c	8414.6 c
LSD 5%	34.04	2.7	611.9

Remark: Numbers accompanied by the same letter in the same column show that they are not significantly different based on the Least Significant Difference (LSD) test with P<0.05. DAP = day after transplanting.

Yield component. The combination treatment between plant spacing and dose of chicken manure gave a significant effect on the number of fruits, fresh weight of fruit, fresh weight of fruit per plant, and fruit yield per hectare. The highest number of fruits in the 40x50 cm spacing + 8 t ha^{-1} treatment of chicken manure was 5.7 pieces.

Table 2. Number of fruit, fruit weight per fruit, fruit weight per plant and yield of sponge gourd.

PlantSpacing (cm)+ Chicken Manure (t ha^{-1})	Number of Fruits (fruit)	Fresh fruit weight per fruit (gfruit^{-1})	Fresh fruit weight per plant (kgplant^{-1})	Yield of Luffa (tha^{-1})
(60x50) + 4	4.3 a	312.0 a	1.3 a	32.4 a
(40x50) + 4	4.3 a	313.0 a	1.3 a	32.6 a
(60x50) + 6	4.5 a	318.0 a	1.4 ab	33.8 ab
(40x50) + 6	4.6 ab	317.0 a	1.5 b	34.9 ab
(60x50) + 8	4.9 b	341.5 b	1.7 c	38.5 b
(40x50) + 8	5.7 c	334.0 b	1.9 d	44.4 c
LSD 5%	0.3	11.3	0.15	5.1

Remark: Numbers accompanied by the same letter in the same column show that they are not significantly different based on the Least Significant Difference (LSD) test with P<0.05.Data were taken from six harvests at the age of 38 to 53 days after transplanting (DAT).

The average number of fruit with a spacing of 60x50 cm + 8 tonha^{-1} of chicken manure did not show a significant difference with a spacing of 40x50 cm + 6 t ha^{-1} of chicken manure, but was significantly higher than the combination of plant spacing and chicken manure with a lower dose (Tabel 2).Application of chicken manure containing sufficient nutrients for plants, especially P (10.21%) will encourage the formation of more flowers and more fruit. This is in line with the opinion of [8], that plants will grow well and produce high yields if the nutrients provided are available in sufficient quantities. Plant spacing also affects the yield of the number of fruits obtained. Plant spacing that is set more densely with sufficient nutrients will produce more fruit.

The combination of spacing and dose of chicken manure gave a significant effect on fresh fruit weight per fruit. Application of chicken manure 8 t ha^{-1}, both at a distance of 60x50 cm and 40x50 cm gave higher fruit weight per fruit than the lower dose of chicken manure. [9], suggested that an increase in plant production can be done by adjusting the spacing of plants but also in combination with the provision of sufficient nutrients, in which the internal competition of plants in the utilization of nutrients, water, and light can be minimized.

Fruit weight per plant also showed the same response trend, the combination treatment between spacing and dose of chicken manure gave a significant effect on fresh fruit weight per plant. The average weight of fresh fruit per plant sample was highest in the 40x50 cm + 8 ton ha^{-1} treatment followed by the 60x50 cm + 8 t ha^{-1} treatment, which was 1.9 kg per plant. While the spacing treatment with chicken manure with a lower dose gave lower fruit weight per plant (Table 2).This is consistent with data on the number of fruit planted and fruit weight per fruit, where the dose of chicken manure up to a dose of 8 tons ha^{-1} resulted in higher fruit weight per plant than the lower dose. This is because chicken manure has a relatively higher P nutrient content than other manures.In addition, the high nutrient content of N (2.11%) in chicken manure stimulates plant growth in general. Nitrogen plays a role in the formation of chlorophyll, amino acids, fats and enzymes. Meanwhile, nutrient P (10.21%) plays a role in root growth and development. Element K (4.9%) helps the formation of proteins and minerals and increases plant resistance to disease [10].

The same trend also occurs in the parameters of fruit yield per hectare. The highest average yield of gambas in the treatment of 40x50 cm + 8 ton ha^{-1} of chicken manure was 44.4 t ha^{-1}. The use of spacing that is too wide is also very inefficient, because the yield per unit area of land obtained decreases with a decrease in the number of plant populations. This is in accordance with the opinion of [11] stated that if the increase in population is still below the increase in competition, the increase in production or yield of crops will be achieved higher at denser spacings.

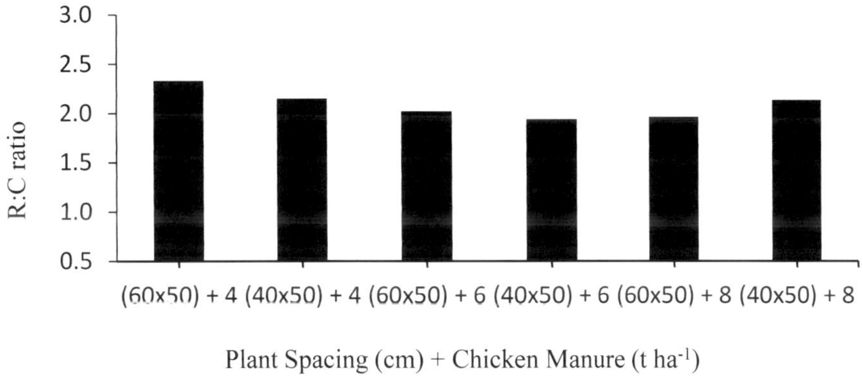

Figure 1. Analysis of R:C ratio of sponge gourd plants in various treatments

Based on the results of research that has been carried out on farm business analysis on spacing and doses of chicken manure, it is stated that the R:C ratio obtained is more than one, which means that farming in the cultivation of gambas can be said to be profitable or feasible to do (Figure 1). The highest R:C ratio was obtained in treatment P0 with an R:C ratio of 2.33 and for treatment P3 had the lowest R:C ratio of 1.94. Each treatment had a different R:C Ratio value, but all treatments had an R:C ratio >1. This is in line with [12], if the R:C ratio > 1 then the carried out earns a profit and is feasible to be cultivated, while if the R:C ratio obtained is <1 then the business carried out is at a loss and is not feasible to be cultivated and if R : C ratio = 1 then the business does not get a profit or break even.

4. Conclusion

Based on the results of the study showed that the treatment with a spacing of 40 × 50 cm + chicken manure (8 t ha^{-1}) had the best results on the growth and harvest parameters of the gambas plant. The treatment with a spacing of 40 × 50 cm + chicken manure (8 tha^{-1}) had a yield of 44.4 t ha^{-1} per ha and an increase of 37% compared to the treatment with a spacing of 60 × 50 cm + chicken manure (4 t ha^{-1}), and an increase of 15% compared to treatment with 60 × 50 cm spacing + chicken manure (8 t ha^{-1}). All treatments are feasible because they have an R:C Ratio of more than 1. The treatment with a spacing of 40 × 50 cm + chicken manure (8 t ha^{-1}) is feasible because it has an R:C ratio of 2.13 and can be used for the growing season. Furthermore, due to the provision of chicken manure on a larger scale.

References

[1] Sumarni, N., Rosliani, R., & Suwandi. 2013. Optimasi Jarak Tanam dan Dosis Pupuk NPK untuk Produksi Bawang Merah dari Benih Umbi Mini di Dataran Tinggi. Jurnal Hortikultura.*22*(2), 148. https://doi.org/10.21082/jhort.v22n2.2012.p148-155

[2] Laude, S., Tambing, Y., Budidaya, J., & Pertanian, F. 2010. PERTUMBUHAN DAN HASIL BAWANG DAUN (*Allium Fistulosum* L.) PADA BERBAGAI DOSIS PUPUK KANDANG AYAM The Growth and Yield of Spring Onion (*Allium Fistulosum* L.) At Various Application of

Chicken Manure Doses. J. Agroland.17(2), 144–148.
[3] Sarido, A. dan La. 2013. Uji Empat Jenis Pupuk Kandang Terhadap Pertumbuhan Dan Hasil Tanaman Cabai Keriting (*Capsicum Annum* L.). Agrifor. 12(1), 22–29. https://www.neliti.com/publications/30076/
[4] El Gendy, S. 2017. Effect of Different Plant Spacing on Growth and Yield of luffa. Journal of Plant Production. 8(4), 485–487. https://doi.org/10.21608/jpp.2017.40015
[5] Shah, P. S., Nabi, G., Khan, M., Hussain, S., & Jan, J. A. 2020. Performance of sponge gourd cultivars under organic and inorganic fertilizer regimes. Sarhad Journal of Agriculture, 36(2), 478–488. https://doi.org/10.17582/JOURNAL.SJA/2020/36.2.478.488
[6] Choudhary, K., Paliwal, R., Mahariya, J. S., Singh, H., & Gupta, N. K. 2020. Plant Growth Improvement in Ridge Gourd through Integrated Nutrient Management Practices (*Xanthomonas axonopodis* pv. punicae). International Journal of Current Microbiology and Applied Sciences.9(3), 1228–1234. https://doi.org/10.20546/ijcmas.2020.903.143
[7] Ramos, D. D., do Carmo Vieira, M., Zárate, N. A. H., Yamamoto, N. T., Carnevali, T. O., & Souza, N. H. 2011. Espaçamentos entre plantas e cama-de-frango na produção de rosela. Acta Scientiarum – Agronomy.33(4), 695–700. https://doi.org/10.4025/actasciagron.v33i4.6933
[8] Hilli, J. S., Vyakarnaal, B. S., Biradar, D. P., & Hunje, R. 2009. Influence of method of trailing and fertilizer levels on seed yield of ridgegourd (*Luffa acutangula* L. Roxb). Karnataka Journal of Agricultural Science. 22(1), 47–52.
[9] Sopha, G. A. 2020. Influence of Plant Density, Compost and Biofertilizer on True Shallot Seed Growth in Alluvial Soil. Indonesian Journal of Agricultural Science. 21(2), 70. https://doi.org/10.21082/ijas.v21n2.2020.p70-77
[10] Alridiwirsah, Rada Mulia Lubis, A. N. 2018.*THE EFFECT OF PRUNING AND CHICKEN MANURE ON VEGETATIVE GROWTH OF HONEY DELI (Syzygium aqueum Burn F.) IN 9 MONTHS AGE*. Icosaanrm.
[11] Kolling, D. F., Sangoi, L., Leolato, L. S., Panison, F., Coelho, A. E., & Kuneski, H. F. 2019. Can an increase in nitrogen rate mitigate damages caused by uneven spatial distribution of maize plants at the sowing row? *Acta Scientiarum* - Agronomy, 41(1). 1–9. https://doi.org/10.4025/actasciagron.v41i1.39874
[12] Dastra, I. G., Martiningsih, N. G. A. G. E., & Sukanteri, P. 2019. Analisis pendapatan usaha tani bunga kembang seribu (Hydrangea) Studi Kasus di Desa Gobleg. Agrimeta. 9(18), 15–18.

IOP Conf. Series: Earth and Environmental Science 1131 (2023) 012016 doi:10.1088/1755-1315/1131/1/012016

The Decreasing Interest of Farmers in Soybean Farming in Puger Sub-District Jember Regency

Nurul Fathiyah Fauzi*, Sisi Dwi Andriyani, Aisyah Yuristianti Utami

Universitas Muhammadiyah Jember, Jember, Indonesia

*nurul.fauzi@unmuhjember.ac.id

Abstract. Puger Subdistrict is a soybean producer in Jember Regency, where the amount of production in the last three years has decreased drastically. In 2017 the soybean production in Puger sub-district reached 53 tons, in 2018 it fell to 508 quintals, and in 2019 it fell to 39 quintals. The decline in soybean production will threaten food security. It is also known that based on statistical data, Puger has decreased the area of soybean harvested, from 201 to 14 hectares ectara during the 2018-2019 period. The objectives of this study are to 1) identify the factors that influence the decrease in farmers 'interest in doing soybean farming in Puger, and 2) provide alternative strategies that can be used to increase farmers' interest in soybean farming in Puger. The research methods are descriptive, Multiple Linier Regression, and Force Field Analysis (FFA) with incidental sampling (45 people and 3 expert respondents). This research used primary and secondary data. The results showed that 1) The factor that had a significant effect on reducing farmers' interest in soybean farming was the price of soybeans (Rp/kilo), while the price of corn (Rp/kilo) and the weather factor did not have a significant effect, 2) the alternative strategies are: increasing the selling price of soybeans at the farmer level, training from extension agents to farmers to increase the ability of farmers in doing soybean farming, especially related to technical matters, and strengthening central government policies stabilization of local soybean prices as an incentive to foster farmers' interest in planting soybeans which will have an impact on increasing the amount of local soybean production that can meet market needs and reduce the number of soybean imports.

1. Introduction

Soybean production in Jember Regency when viewed from the data for the last 3 years has decreased. Production data in 2017 was 19,164 tons, while in the following year it continued to decline to 13,886 quintals in 2018 and 12,521 quintals in 2019 (1). Puger sub-district is one of the soybean-producing sub-districts, it's just that the amount of production in the last three years has decreased very drastically. In 2017 the soybean production in Puger sub-district reached 53 tons, in 2018 it fell to 508 quintals, and in 2019 it fell to 39 quintals. The decline in soybean production will threaten food security. It is also known that based on BPS data from Puger Subdistrict, there was a decrease in soybean harvested area, from 201 Ha to 14 Ha during the 2018-2019 period [1].

This symptom of a decline in soybean production in the Jember Regency in general and Puger, in particular, is an indication that farmers are starting to experience a decline in interest in not doing soybean farming. Many factors are suspected to be the trigger for the decline in farmers' interest in soybean farming, it's like the soybean price factor. This condition is an interesting phenomenon to be studied further and find a solution in the form of an appropriate alternative strategy to overcome the decline in production.

The objectives to be achieved from this research are: 1) Identifying the factors that influence the decrease in farmers' interest in doing soybean farming in Puger, and 2) Providing alternative strategies to increasing farmers' interest in soybean farming in Puger.

2. Material and Methods

2.1. Method of Study

Study used descriptive and analytic method.This methodexamine status of respondent, object, set of condition, system of thinking, condition or occurance in once time, testing hypothesis and interpreting analysis result deeply.

2.2. Field of Study

Field of study was determined by purposive methodin Puger sub-district because in Puger there is a decrease in the amount of soybean production and planting area from year to year.

2.3. Methodof Sampling

Method of sampling by incidental sampling. The number of samples taken for Questionnaire I was 45 respondents, while for Questionnaire II with an expert sample of 3 people (2 respondents from the expert soybean farmer and 1 respondent from the Department of Agriculture, Food Crops and Horticulture, Jember Regency).

2.4. Methodof Data Analysis

Analysis of the data to answer the first objective with Multiple Linear Regression Analysis. The data to be tested can be called good data if it meets the BLUE (Best Linear Unbiased Estimator). The BLUE test including 1) Normality, 2) Autocorrelation, 3) Multicollinearity, and 4) Heteroscedasticity [2]. This research didn't use an autocorrelation test because the data used in the study is cross-section data, while the error rate used is 5%. The equation model:

$$Y= \beta_0 + \beta_1X_1 + \beta_2X_2 + \beta_3X_3 + e$$

Explanation:

Y	= The interest of farmers
$\beta_0 .. \beta_x$	= constanta
X_1	= price of soybean (Rp/kilo)
X_2	= price of corn (Rp/kilo)
X_3	= weather (Likert scale method)
e	= natural logarithm = 2,7183

To test how far the Y is caused by variations in the X variables the Coefficient of Determination or adjusted R Square (R2) is used. If the adjusted R Square value is greater, it indicates that there is a very strong relationship between the X and the Y variable. Then test the hypothesis that has been proposed using statistical tests:

a) The ANOVA test criteria:

a. Sig F > Sig α (0,05) means that the independent variables jointly affect the dependent variable.

b. Sig F $\leq$ Sig α (0,05) means that the independent variables together do not affect the dependent variable.

b) Partially test criteria:

a. If Sig $\leq \alpha$ (0,05) means that the Xi variable has a significant effect on the Y variable.

b. If Sig > α (0,05) means that the Xi variable has no significant effect on the Y variable.

The analytical method to answer the second objective is used Force Field Analysis (FFA). The stages in conducting the FFA analysis are as follows [3]:

1. Identify problems based on strategic issues.
2. Analyze the problem by identifying various driving and restraining forces.
3. Provide a priority scale assessment of each driving and resistor factor.

The assessment of each identified factor will determine the success factor of the goal. There are several aspects that need to be considered in assessing each factor, namely (3):

1. The urgency of the factor to the goal, consisting of the value of urgency (NU) and the weight of the factor (BF).
2. Factor support for goals, consisting of Support Value (ND) and Support Weight Value (NBD).
3. The relations between factors to the goal, consisting of the Correlation Value (NK), the Average Correlation Value (NRK), and the Weighted Linkage Value (NBK).

The NU, ND, and NK assessments use 1-5 scale:a) 5: very high value of urgency; b) 4: a high value of urgency; c) 3: quite high of urgency; d)2: less value of urgency, and e)1: very less value of urgency.

The driving factors came from strengths and opportunities, while the resistor factors from weaknesses and threats. In general, the assessment can be carried out using qualitative values that are quantified using 1-5 scale. Determining the Urgency Value (NU) aspect of each driving and resistor factor by using a comparison technique. To determine the value of the aspect of urgency (NU) on the resistor factors, it is the same as in determining the NU on the driving factors. Furthermore, the NU and BF of each driving and resistor factors are entered into their respective columns (Table 1).

Table 1. Evaluation of Driving and Resistor Factors

No.	Driving &Resistor Factors	NU	BF (%)	ND	NBD	NK						NRK	NBK	TNB	FKK
						D1	D2	D…	H1	H2	H…				
D1															
D2															
D…															
H1															
H2															
H…															

Source: *Sianipar dan Entang, 2003. [3]*

The NBD value is ND x BF
The NRK value is from TNK/(N-1), where TNK is the sum of the correlation values of one factor and N is the number of driving and resistor factors assessed.
The NBK value is obtained from NRK x BF
TNB value is NBD + NBK

To assess the driving and resistor factors, the Key Success Factors (FKK) are used. The determination is as follows (3):
1. Select each resistor factor and driving factor based on the largest TNB or Total Factor Weight Value.
2. If TNB is the same, then the largest BF is chosen.
3. If BF is the same, then the largest NBD is selected.
4. If the NBD is the same, then the largest NBK is selected.
5. If the NBK is the same, then choose based on experience and rational considerations.

3. Result And Discussion

3.1 The Decreasing of Farmers Interest in Soybean Farming Factors

a. Normality Test

The Skewness Ratio shows a negative result of -1.867, while the Kurtosis Ratio shows the result of 0,525. The value of the Skewness and the Kurtosis ratio is between -2 to +2, so the data used has been shown to be normally distributed.

b. Multicollinearity Test

The results of the analysis show that the VIF value for each independent variablesare less than 10 (in the value of 1.005 - 1.043). Thus, it can be concluded that all independent variables did not experience multicollinearity disorders.

c. Heteroscedasticity Test

The results of the analysis (Glejser test) in the Significance column (Sig.) of all independent variables, the significance value is above 0.05 (Sig > 0.05). This shows that all independent variables are homoscedasticity.The result of Multiple Linear Regression shows as follows:

Table 2. The results of multiple regression analysis (α = 5%)

Model		Unstandardized Coefficients B	Std. Error	Standardized Coefficients Beta	T	Sig.
1	(Constant)	2.580	.896		2.878	.006
	X1	9.024	.000	.160	.021	.013
	X2	-4.624	.000	-.110	-.703	.486
	X3	-.048	.152	-.048	-.312	.757

Sig ANOVA: 0,046
R Square: 0,723
Adj R Square: 0,698

The Adj R Square is 0.698 or 69.8% the decrease in interest in farmers in conducting farming soybean is influenced by all the third independent variables, while 30.2% influenced factor, availability of production facilities, and other factors. Meanwhile, if viewed from the ANOVA significance is 0.046 it shows that simultaneously, the three independent variables had a significant effect. The equation for the multiple linear regression function from this research is:

$$Y= 2,580 + 9,024X_1 - 4,624X_2 - 0,048X_3 + e$$
$$(0,013) \quad (0,486) \quad (0,757)$$

a. The value of the constant on the model is 2,580, if there is no effect of all the independent variables then the interest of farmers in soybean farming is only 2,580 units.

b. The X_1coefficient is 9,024, means that every time there is a decrease in the price of soybeans of Rp. 1000/kilo means that it can reduce farmer's interest in soybean farming by 9024 units with other factors which have influence are considered. This condition has a significant effect as seen from the significance value of 0.013. It cannot be denied that the selling price has become an attraction for farmers. If the selling price of soybeans is getting better, this will be a separate stimulus for farmers to continue to do soybean farming.

c. The X_2 coefficient is 4,624 (negative). which means that every time there is an increase in the price of corn by Rp 1000/kiloi.e. it will decrease the interest of farmer for planting soybeans by 4624 units, although this condition does not indicate this significance value of 0,40, which is a significant value of influence. This fact shows that farmers have an interest to switch the corn farming if the price of corn is higher than the price of soybean.

d. The X_3 coefficient is 0.048 (negative), means that every time there is a change in the weather, the weather is not erratic, so it will be able to reduce the interest of farmers in doing soybean farming. If we look at the statistical results, it shows that this does not have a significant effect (0.757 > 0.05). From the results of observations, even though there is news regarding the weather conditions, it is still raining continuously during the planting until the harvest periodof soybeans. In fact, when the soybean harvest season arrives, the rain falls almost every day and destroys the harvest yields.

3.2. Alternative Strategies in Efforts to Increase Farmers Interest in Soybean Farming

The identification of the driving and resistor factors that has been carried out by the FFA analysis shows the following results.

Table 3. Identification of the drivers and resistors of the strategy in increasing farmers' interest in soybean farming in puger

No	Driving Factors	TNB	FKK	No	Resistor Factors	TNB	FFK
	Strengths				*Weaknesses*		
D1	Low soybean price	2,58	*	H1	Farming costs are relatively cheaper than other farming in the 3rd season (corn)	1,33	
D2	No subsidies	0,19					
				H2	Availability of production inputs can be obtained anywhere	1,45	*
Total *Strengths*		2,77		Total *Weaknesses*		2,78	
	Opportunities				*Threats*		
D3	Crop failure due to pest and disease attacks	0,93		H3	Many seeds of superior varieties	0,44	
D4	Weather does not support (Continuous rain)	1,98	*	H4	Market needs	3,03	*
D5	The selling price of subsidized commodity (corn) is higher than soybean	1,50		H5	There is an extension of the soybean farming	0,56	
Total *Opportunities*		4,41		Total *Threats*		4,03	
Total Driving Factors		7,18		Total Resistor Factors		6,81	

Note: *) The main factors.

The alternative strategies to increase farmers' interest in doing soybean farming in Puger are:

1. Increasing the selling price of soybeans at the farmer level is the main key in returning farmers' interest to plant soybeans.
2. Coaching of extension workers to farmers to improve the ability of farmers in soybean farming, especially related to technical matters. This is mainly related to the handling of soybean farming when the weather is uncertain
3. Strengthening of central government policies through stabilizing local soybean prices as a stimulus to foster interest in farmers planting soybeans which will have an impact on increasing the amount of local soybean production that is able to meet market needs and reduce the amount of soybean imports.

4. Conclusion

1. Factors that significantly influence farmers' interest in soybean farming are the price of soybeans, while the price of corn and weather factors do not have a significant effect.
2. Alternative strategies to increase farmers' interest in doing soybean farming in Puger are:
   a. Increasing the selling price of soybeans at the farmer level is the main key in returning farmers' interest in growing soybeans.
   b. Coaching of extension workers to farmers to improve the ability of farmers in soybean farming, especially related to technical matters. This is mainly related to the handling of soybean farming when the weather is uncertain

c. Strengthening of central government policies through stabilizing local soybean prices as a stimulus to foster interest in farmers planting soybeans which will have an impact on increasing the amount of local soybean production that is able to meet market needs and reduce the amount of soybean imports.

References

[1] Badan Pusat StatistikKabupatenJember. 2020. *KabupatenJemberDalam Angka 2020.* https://jemberkab.bps.go.id/.

[2] Setyadharma, A. 2010. *Uji Asumsi Klasik Dengan SPSS 16.0.* Fakultas Ekonomi Universitas Negeri Semarang.

[3] Sianipar dan Entang. 2003. *Teknik-Teknik Analisis Maanajemen.* Jakarta: Lembaga Administrasi Negara.

Study of antibiosis of *Streptomyces* sp. from the land of shallot plants as biological agents of *Fusarium*sp. cause of Twisted diseases (Moler)

Penta Suryaminarsih[1]*, Tri Mujoko[1], K Gusriyan[1], Fitri Wijayanti[1], Salmah Mohammad[2]

[1]Universitas Pembangunan Nasional "Veteran" Jawa Timur, Indonesia
[2]Universiti of Sultan Zainudin Malaysia, Malaysia

*penta_s@upnjatim.ac.id

Abstract. The use of bio pesticides and organic pesticides from secondary metabolic filtrate of APH microorganisms is one of the food and horticulture cultivation technology packages which are the basis for this research objective. The research objective was to explore , isolate and identify of Streptomyces sp. in the land of shallot plants (Abm) and the antibiosis against the pathogens Fusariumsp cause of Moler shallot plant in vitro. Observer of Quality and quantity characteristics of secondary metabolites as antibiosis, The results showed that Streptomyces sp from land of shallot plant (Abm) isolates producing little antibiosis less able to inhibit the development of the pathogens Fusariumsp cause of Moler shallot plant in vitro. The identification of Actinomycetes spp. based on the morphology characteristic showed that Actinomycetes isolates are closely related with Streptomyces sp.

1. Introduction

The prospect of developing onion production in Indonesia ranks first in ASEAN Countries, and experienced a 3.70% increase in harvest area growth in 2010-2014 compared to the previous year [1]. However, until now it is still unable to meet domestic production, causing fluctuating price volatility and contributing to inflation from the agricultural sector [2]. Some of the factors causing low productivity include low soil fertility rates, increased attacks of plant disrupting organisms (OPT), microclimate changes and low-grade seedlings [3]

The world market for organic agricultural products is increasing by 20% per year, therefore the development of organic agricultural cultivation needs to prioritize high-value crops for the domestic and export markets. A sustainable agricultural system based on organic agriculture is one of the world's needs for Fresh Plant-Originated Products (PSAT) that are safe and environmentally friendly. The use of microorganisms in sustainable agriculture is one of the nanotechnology innovations for the control of Plant Pest Organisms (OPT) that are safe, environmentally friendly and sustainable [4].

The development of shallot production in Indonesia ranks first in ASEAN countries and increas in the growth of harvested area by 3.70% in 2010-2014 [1]. This causes country fluctuating price and contributing to inflation from the agricultural sector [5]. Several factors cause low productivity include low soil fertility, an increase in attack by plant pest organisms (OPT), changes in the microclimate and low quality seeds used [6].

Plant disease pests are one of the factors that cause the decline in the production of hicortultural plants onion (Allium cepa L). Once of pathogens that attack shallot plants is a wilt disease caused by Fusarium sp. The pathogen *Fusariumcapsici* causes the base of the tuber to rot, the leeks turn yellow and become twisted (moler) and the plant is easily uprooted. Infected tubers will show a white tuber base because the mass of fungi and rotting tubers starts from the tuber base [5].Controlling using synthetic chemicals can increase environmental pollution and plant foodstuffs [7]. Therefore, environmentally friendly control is needed by utilizing microorganisms as natural enemies. Several research results indicate that *Streptomyces*spp can be used as a biological agent and also acts as PGP. Streptomyces griseurubens can control fusarium wilt and increase the growth of tomato plants [8]. The results of a study by Gopalakrishnan et al, 2021 [9]indicated that the selected strains of *Streptomyces* spp. have the potential for biological control of charcoal rot disease in sorghum.

Secondary metabolites are organic compounds that are not directly involved in the normal growth, development and reproduction of organisms and are formed during the final or nearing the stationary stage of growth [10]. Research on *Streptomyces* spp. has been carried out by each member of the researcher as well as joint research. The results showed that *Streptomyces* spp. which has been studied, namely *S griseus* which comes from chili fields in the Pare area increases the growth and production of tomatoes and is able to control the intensity offusarium wilt disease (8). The antagonist mechanism used by Streptomyces sp. include Antibiosis by issuing various kinds of antibiotics. One of the antibiotics produced by S. griseus that was discovered and identified by [11] is actidione which has the chemical formula $C16H22NO4$.

Antibiosis is once of mechanism of Biological Control Agents (BCAs) and secondary metabolites are very important innovations to know, so that the control is carried out successfully. This research can provide information on the success of BCAs and secondary metabolite filtrate formulation of *Streptomyces* sp. and needs to be investigated.

2. Materials and methods

Screening, isolation and identification of BCAs are carried out by survey and description methods. Test antagonist BCAs *Streptomyces* sp. using the Honest Difference Test (BNJ) between two treatments i.e. control without BCAs and using BCAs *Streptomyces* sp. compiled in Complete Random Design (CRD). The antagonist test uses of concentrated difference secondary metabolite crude extract. Treatment arranged in a complete randomized design with treatments of 5%, 10%, 15% and 20%. All treatments of APH and secondary metabolites are repeated 4 times each.

*2.1 Isolation and identification of*Streptomyces sp.

Isolation of biological agents microbes, using the soil plating method by Dhingra and Sinclair (Singleton, Michail, and Rush, 1993): Soil from shallot fields was weighed 1 gram by analytical scales, then make a suspension 10^{-4} dilution. 1 mL of dilution was taken aseptically inserted and flattened into the GNA medium in a Petri dish for the isolation of *Streptomyces* sp. The biological agents were obtained, then purified, and reproduced on PDA media in Petri dishes. *Streptomyces* sp. isolates were observed the colony morphology and spore morphology by microscopically observed

2.2 Isolation and identification of Pathogens

The method used for the isolation of pathogens was the stimulation of spores from fresh materials (Sastrahidayat, 1994), the pathogen was isolated from shallot plants from the field (Pare-Kediri District) which were attacked by *Fusarium* sp.. The diseased plant stems was cleaned, sterilized it with 70% alcohol, air dried, then slice it with a scalpel. The incision was inoculated on the PDA medium. The pathogenic fungi that grew, were isolated and purified, then the pure

*Fusarium*sp fungi were propagated on PDA media. Koch's postulate test was carried out to ensure that the isolation results were the cause of Moler disease

2.3 Production of secondary metabolites

Isolate *Streptomyces sp.* is inoculated in 150 ml of sugar potato extract (EKG) in a 150 ml Erlenmeyer 250 ml (Harni, 2017) Erlenmeyer 1 is filled with 6 pieces of agar containing a single isolate of Streptomyces sp. (diameter ± 8mm), Erlenmeyer 2 is filled with 6 pieces of agar. Secondary metabolite extraction was carried out by separating the secondary metabolite compounds Trichoderma sp. and Streptomyces sp. from the media and cells of Streptomyces sp. with a speed of 4,000 rpm for 15 minutes (Harni, 2017). Then filter it using Whatman paper no. 1. To get a more sterile result, the supernatant was filtered again using a 0.2 µm siringefilter which was then used for further testing.

2.4 Antagonism in vitro test

Antagonist test is a preliminary test conducted to determine the biological agents against fungal pathogens. The media used in the antagonist test was PDA media with the dual culture method. The 7-day-old culture of *Fusarium* sp. was cut using a 5 mm diameter cork borer and placed opposite the *Streptomyces* sp. 14 days old. Then it was incubated and its development was observed for 7 days.:

$$DI = \frac{Dc - Dt}{Dt} \times 100$$

DI is the percentage inhibition;
Dc is the diameter of the control;
Dc is the diameter of inhibition.

Antibiosis test consist of The research method used was a completely randomized design (CRD) with 4 treatments, each treatment was repeated 4 times which was then further tested using BNJ with a 5% confidence level. Treatment used were :without using secondary metabolites (P0), 10% concentration of secondary metabolites (P1), 20% concentration of secondary metabolites (P2), secondary metabolite concentration 30% (P3)

3. Results and discussion

3.1 Isolation and identification of Streptomyces sp.

The results of isolation of Streptomyces in the shallot field, found one isolates which had colony morphology and smel characteristics such as Streptomyces. Colonies are cream in color, the surface is slightly rough and dense. Microscopic observations showed the presence of condensed hyphae and spores arranged in chains.Identification of actinomycetesof dickey zeae pineapple shown the isolate like Streptomyces is also known for producing volatile compounds, namely Geosmin which has a distinctive smell on the soil.Streptomyces belongs to the Actinomyces group, namely bacteria that have a branched hyphae structure resembling a fungus and can produce spores[12])

3.2. *Isolation and identification of Pathogens*

AThe results of pathogen isolation from Moler Shalot deseases were characterized by white colonies with spreading margins and fibrous mycelium like cotton. Microconidia observation showed that the culture had single-celled spherical and crescent-shaped macroconidia with 3-5 septa (Fig. 1). In Accordance with [13] research , the Colony of Fusarium cause of shallot diseases found had white, cream and purplish colonies, almost of macroconidia have crescent-shaped with 3-5 septa and microconidia oval in shape, long, slender middle and blunt ends.

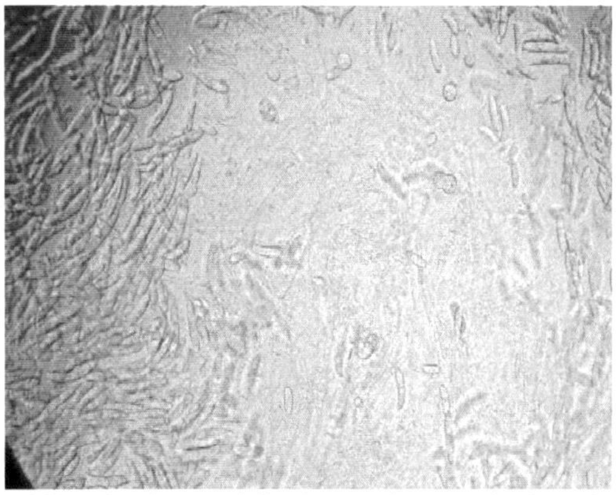

Figure 1. Macroconidi and microconidi of Fusarium sp. isolate from twisted, wilting shallot plant diseases

The Koch postulate test of *Fusarium* sp. pathogenicity showed plant be wilting, Leaves turn yellow and curl, Easy to remove , Bulbs and roots rot. The wilting symptom of shallot could be caused by *F. solani* and *F. acutatum*, while bulb rot could be caused by *F. solani*, *F. acutatum* or *F. oxysporum*, and twisted disease could be caused by *F. solani* or *F. acutatum*[14].

3.3. Antagonism in vitro test

Antagonist test is a preliminary test conducted to determine the biological agents against fungal pathogens. Antagonist test of *Streptomyces* sp. against isolates of the fungal pathogen *Fusarium* sp. using the Honest Difference Test (BNJ) showed the average percentage of inhibition ability was 34%. *Streptomyces* sp. can inhibit the development of *Fusarium* sp. due to the content of secondary metabolite antibiotics. The inhibition of secondary metabolites of *Streptomyces*sp against *Fusarium*sp using BNJ with a confidence level of 5% showed that the inhibitory of secondary metabolites with concentrations of 10% and 20% did not differ in effect, able to inhibit 30% - 36.5%, while the 30% concentration was able to inhibit up to 50% (Fig 2)

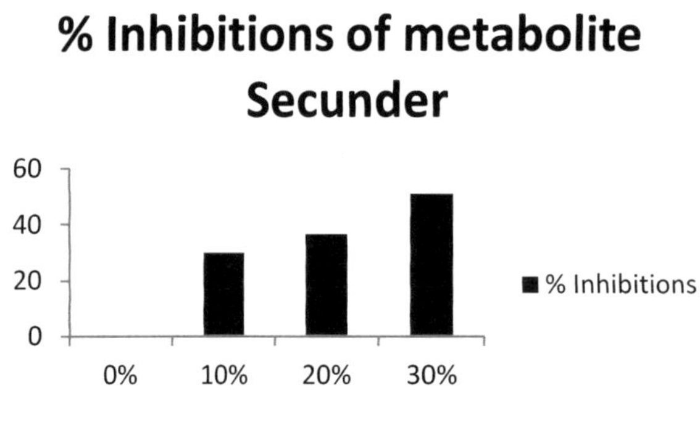

Figure 2. Diagram of the average inhibition of secondary metabolites crude extract of *Streptomyces* sp. against *Fusarium* sp.

The results of microscopic observations showed that the higher concentration of the secondary metabolite extract of Streptomyces sp. causes the formation of spores and the diameter of the hyphae was getting smaller and lysis and chlamydosporeswere formed. In the treatment without *Streptomyces* sp. Microconidiawere formed, whereas at concentrations of 10% and 20%, very little microconidia were formed.In vitro bioassay studies, Peng et al. 2020 [15] revealed that FC strongly inhibited mycelial growth and conidia germination in *Fusariumoxysporum*. The results showed that some *Streptomyces* spp. which has been studied wereable to control the intensity offusarium wilt disease [9]. The antagonist mechanism used by *Streptomyces sp.* include Antibiosis by issuing various kinds of antibiotics.

4. Conclusions

The results showed that Streptomyces sp from land of shallot plant (Abm) isolates producing little antibiosis less able to inhibit the development of the pathogens Fusariumsp cause of Moler shallot plant in vitro 30 % - 51 %. The identification of Actinomycetes sp. based on the morphology characteristic showed that Actinomycetes isolates are closely related with Streptomyces sp.

Knowledge
This article was compiled because of the first year UPN "East Java Veterans" internal grant. for the second year the research is continued in collaboration with UNSZA Malaysia and UNTAD Palu, thank you

References

[1]Badan Pusat Statistik. (2020). *Produksi Tanaman Sayuran*. Jakarta

[2]Setiawan, A. F., &Hadianto, A. (2014). Fluktuasi harga komoditas pangan dan dampaknya terhadap inflasi di Provinsi Banten. *Journal of Agriculture, Resource and Environmental Economics*, *1*(2), 81-97.

[3]Triharyanto, E., Purnomo, D., &Yunus, A. (2020).Detection of Flowering Ability on Several Bulbs Shallot Sources by using Hd3a and Endogenous GA 3 Analysis. *Indian Journal of Agricultural Research*, *54*(6).

[4]Pasandaran, E., Syakir, M., Heriawan, R., &Yufdy, M. P. (2017).MenujuPertanian Modern Berkelanjutan.

[5]Setiawan, A. F., &Hadianto, A. (2014). Fluktuasi harga komoditas pangan dan dampaknya terhadap inflasi di Provinsi Banten. *Journal of Agriculture, Resource and Environmental Economics*, *1*(2), 81-97.

[6]Triharyanto, E., Nyoto, S., &Yusrifani, I. (2018, March). Aplication of giberelins on flowering and yield of two varieties of shallot in lowland. In *IOP Conference Series: Earth and Environmental Science* (Vol. 142, No. 1, p. 012066).IOP Publishing.

[7]Triwidodo, H., &Tanjung, M. H. (2020).Hama Penyakit Utama Tanaman Bawang Merah (Allium Ascalonicum) dan Tindakan Pengendalian di Brebes, Jawa Tengah. *Agrovigor: JurnalAgroekoteknologi*, *13*(2), 149-154.

[8] Suryaminarsih, P., Harijani, W.S., Muljani, I.R., Mindari, W. & Rahmadhini, N., (2020). Screening and identification of Actinomycetes produced chitinolytic from suppression soil as biological agents of fruit flies (*Bactrocera* sp.). Eurasian *Journal of Biosciences,* 14 (1), 977-982.

[9] Gopalakrishnan, S., Srinivas, V., Naresh, N., Pratyusha, S., Ankati, S., Madhuprakash, J., ... & Sharma, R. (2021). Deciphering the antagonistic effect of Streptomyces spp. and host-plant resistance induction against charcoal rot of sorghum. *Planta*, *253*(2), 1-12.

[10] Santoso S.E. , L Soesanto, danHaryanto, T. (2007). Penekanan Hayati Penyakit Moler Pada Bawang Merah dengan *Trichodermaharzianum*, *Trichodermakoningii*, dan *Pseudomonas fluorescens* P60, *Jurnal Hama danPenyakitTumbuhanTropika,* Vol 7, No 1.

[11] Aeny, T. N., Prasetyo, J., Suharjo, R., Dirmawati, S. R., Efri, E., &Niswati, A. (2018). Isolation and identification of actinomycetes potential as the antagonist of Dickeyazeae pineapple soft rot in Lampung, Indonesia. *Biodiversitas Journal of Biological Diversity*, *19*(6), 2052-2058.

[12]Hikmahwati, H., Auliah, M. R., Ramlah, R., & Fitrianti, F. (2020).IdentifikasiCendawan Penyebab Penyakit Moler Pada Tanaman Bawang Merah (*Allium Ascolonicum L.*) Di Kabupaten Enrekang. *AGROVITAL: Jurnal Ilmu Pertanian*, *5*(2), 83-86.

[13]Lestiyani, A., Wibowo, A., Subandiyah, S., Gambley, C., Ito, S., & Harper, S. (2014, August). Identification of Fusarium spp., the causal agent of twisted disease of shallot. In *XXIX International Horticultural Congress on Horticulture: Sustaining Lives, Livelihoods and Landscapes (IHC2014): 1128* (pp. 155-160).

[14] Peng, C., An, D., Ding, W. X., Zhu, Y. X., Ye, L., & Li, J. (2020). Fungichrom in production by Streptomyces sp. WP-1, an endophyte from Pinusdabeshanensis, and its antifungal activity against *Fusarium oxysporum*. *Applied Microbiology and Biotechnology*, *104*(24), 10437-10449.

IOP Conf. Series: Earth and Environmental Science 1131 (2023) 012018 doi:10.1088/1755-1315/1131/1/012018

Evaluation of the Success of Forest Land Based on Geographic Information Systems, in Wonosalam District, Jombang Regency

Maroeto*, Rossyda P., Mohammad Idhom, Wahyu Santoso, Siswanto

UPN "Veteran" Jawa Timur, Surabaya, Indonesia

*maroeto@upnjatim.ac.id

Abstract. The benefits of forests for life are not only for local communities living around forests, but also for global communities whose lives depend on forests as a resource that is able to provide various functions related to environmental management. The research aims to identify the potential of the Wonosalam forest area planned for UPN Integrated Farming & Forestry Tourism (UIFFT) - KHDTK Wonosalam in support of conservation policies by mapping forest areas based on land suitability characteristics and land availability suitable for KHDTK development. This type of research is survey research. and spatial analysis using Geographic Information Systems. Location in this research was conducted in the Wonosalam Forest Area in Wonosalam District. Wonosalam District is located in Jombang Regency with an area of 121.63 km^2. The results of the study can be concluded that 1. the results of the survey of regional characteristics and physical and chemical analysis of the soil for durian, clove and coffee plants have land suitability classes that are actually in accordance with marginal (S3eh) with a limiting factor on erosion hazard while for potential after experiencing improvements that are not overall permanent value obtained is quite appropriate (S2) with various limiting factors but most of the constraints on erosion hazard and land preparation. 2. The direction of forest area development in future research areas based on land suitability is an integrated farming forestry area with a technological approach, and plant suitability so that this area can become a center for research, innovation, education, training, empowerment and community assistance without compromising its function. forests as environmental conservationists in an integrated manner.

1. Introduction

Forests are one of the natural resources that are priceless and can provide great benefits for human welfare, both tangible benefits that are felt directly and intangible benefits that are felt indirectly. The management of forest resources is inseparable from the legal basis of Article 6 of Law no. 41 of 1999 (on Forestry) which states that efforts to manage natural resources in forest areas are carried out through protection, conservation and production functions by taking into account the continuity of their supply and the surrounding environment.

The management of forest resources cannot be separated from the legal basis of Article 6 of Law no. 41 of 1999 (on Forestry) which states that efforts to manage natural resources in forest areas are carried out through protection, conservation and production functions by taking into account the continuity of their supply and the surrounding environment. The goal is to strive for the preservation of forest resources and ecosystem balance, so that they can better support efforts to improve community welfare and environmental existence (Datta, 2014; Van Noordwijk and Hairiah, 2006). Specifically for the protection function, the government has pursued Law 32 of 2009 (on Environmental Protection and Management) which mandates that environmental protection and management is a systematic and integrated effort carried out to preserve environmental functions and prevent pollution and/or environmental damage which includes planning, utilization, control, maintenance (Rossiter, 1994) supervision, and law enforcement.The consequences clearly have an impact on the existence of the area, including the presence of the community around the area, as happened in the Wonosalam forest area in Jombang Regency.

The reality of various problems related to forest management by the government is then formulated a policy, namely through the Regulation of the Minister of Environment and Forestry of the Republic of Indonesia, forest areas which is specifically intended for the interests of forestry research and development, forestry education and training as well as religion and culture.

This study intends to identify the potential of the Wonosalam forest area planned for UPN Integrated Farming & Forestry Tourism (UIFFT) - KHDTK Wonosalam in support of conservation policies by mapping forest areas based on land suitability characteristics for durian, clove and coffee plants as well as the availability of suitable land for development. KHDTK Wonosalam.

2. Material and Methods

2.1. Research Design

This research design is designed to answer questions that have been formulated and the objectives to be achieved, through hypothesis testing. This type of research is survey research. surveys can also be considered as explanatory research which intends to explain the relationship between variables through testing the observed hypotheses (Singarimbun and Effendi, 2000).

The research approach used in this research is quantitative research and spatial analysis using geographic information systems. The initial step of this research was built based on the creation of a spatial map that helped when starting a study to identify potential physical resources of forest land based on land suitability suitable for agroforestry models, namely durian, clove and coffee plants. The research stages include data collection, image interpretation and field verification.

2.2. Location

Location in this study This research was carried out in the Wonosalam Forest Area in Wonosalam District. Wonosalam District is located in Jombang Regency, Wonosalam District is located at 112 21′ 05″ to 112 23′ 22″ east longitude and 07 44′ 59″s.d 07 40 '01″ south latitude.

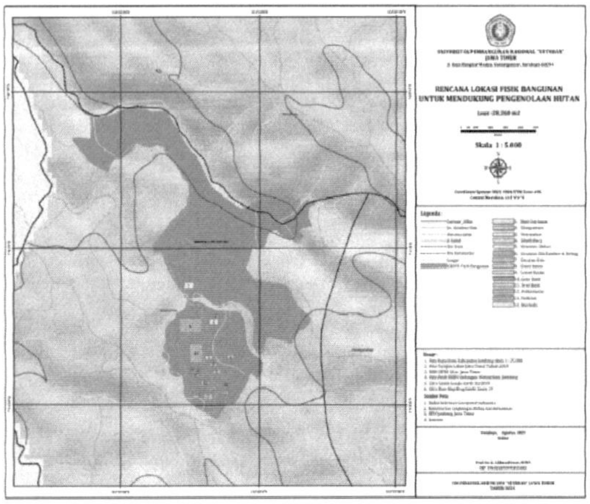

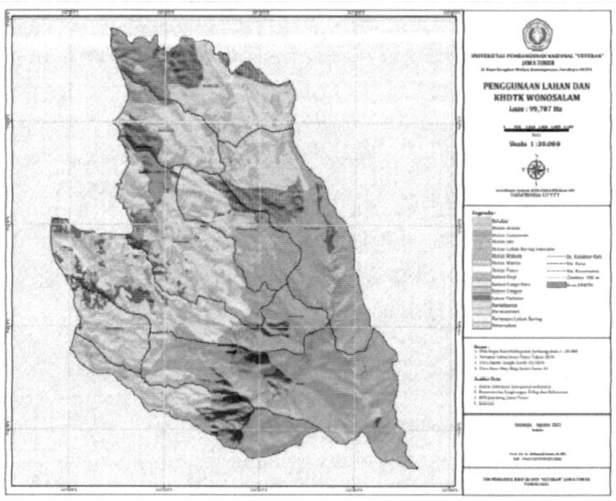

Figure 1. Research Locations**Figure 2.** Land Use Map

2.3. Research Stages

a. Preparation,
b. Ground Check and Sampling

Sampling based on 4 Land Use Units includes
Land Use Units1: Mixed Forest
Land Use Units 2: Pine Forest and Coffee
Land Use Units 3: Pine Forest
Land Use Units 4: Open Land

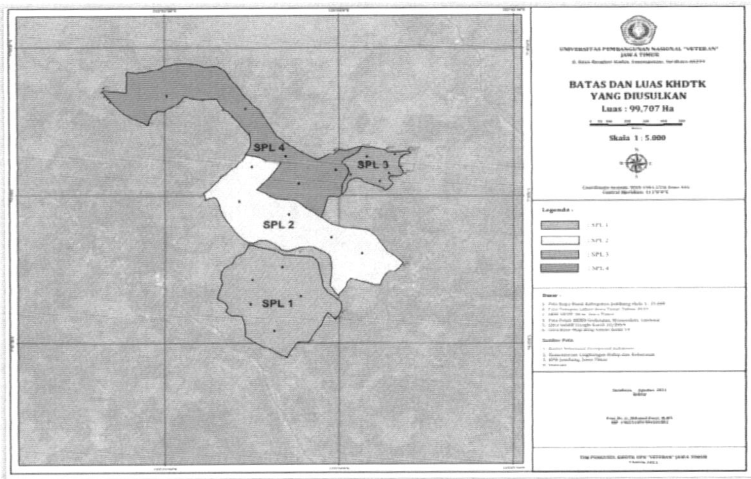

Figure 3. Location of Land Use Units

c. Soil Chemical Properties Analysis

Soil samples collected in the field were air-dried before being tested in the lab. Regional climatological data, as well as morphological, physical, and chemical qualities of the soil, were among the parameters observed.

2.4. Data Analysis Method

According to FAO (1983) In principle, plant needs or plant growth requirements are combined with land features to determine land suitability. As a result, this categorization is also known as species matching.The land suitability class is divided into four levels, namely: very suitable (S1), suitable (S2), marginally suitable (S3) and not suitable (N). The type of barrier is also reflected in the sub-class in this land suitability categorization.There are seven types of inhibitors known, namely e (erosion), w (drainage), s (soil), a (acidity), g (slope) to (soil depth) and c (climate).

3. Results and Discussion

Based on the results of observations in the field and the results of laboratory analysis that evaluation of land suitability is very necessary for land use planning according to the characteristics of land conditions so that the actual and potential forest land suitability class is obtained for the preservation of the forest environment in the highlands region. The results of the description of the actual land suitability class, it can be seen that there are several limiting factors for the actual land suitability class and the potential land suitability class is the land suitability that will be achieved after land improvement efforts are made after being given input according to the level of management that will be applied, it is hoped that land productivity will increase in the future (Charnsungnern and Sittichai, 2017). Based on the results of laboratory analysis for forest areas, it can be described in Table 1.

The suitability of land in various land use units has different limiting factors for certain plant commodities such as durian, cloves, and coffee which are adapted to the wishes of the people in the Wonosalam plains without reducing the function of land for sustainable productive forests. For plantation crops such as durian, cloves, and coffee, these are the main commodities desired by the community around the forest in the Wonosalam region.

In forest land, most of the land in the highlands, especially land use units 1, 2, 3 and 4, especially durian and arabica coffee, are the dominant plants in the area besides pine. The Wonosalam area mostly has alfisol soil types, mountainous and hilly physiography, as well as volcanic ash as parent material. Alfisol soil type has a good enough ability to cultivate plantation crops and production forests.

Table 1. Land Suitability Based on Selected Commodities

Land Condition	**Land Suitability for Wonosalam Area**					
	Clove		**Coffee**		**Durian**	
	Actual	**Potential**	**Actual**	**Potential**	**Actual**	**Potential**
Land Use Units (LUU) 1	S3eh	S2wa,eh	S3eh	S2eh	S3eh	S2eh
Land Use Units(LUU) 2	S3eh	S2wa,eh,lp	S3eh	S2eh,lp	S3eh	S2nr,eh,lp
Land Use Units(LUU) 3	S3eh	S2wa,rc,eh,lp	S3eh	S2rc,nr,eh,lp	S3eh	S3rc,nr,eh,lp
Land Use Units(LUU) 4	S3eh	S2wa,eh,lp	S3eh	S2nr,eh,lp	S3eh	S2nr,eh,lp

Source: Data Analyzed (2017).

Explanation : LUU 1 : Pine forest and coffee tc : Temperature nr : Nutrient retention
LUU2:Mixed pine forest wa : Availability of water eh : Erosion hazard
LUU3:Dense pine forest rc : Root media lp : Land preparation
LUU4:Open land

Basically alfisol can be used for cultivation. But because it has an extreme steep slope, this area only needs to be planted with hard plants such as teak, pine or fir trees to anticipate heavy erosion but can also be intercropped with plantation crops. The moderate to highlands of the study area are in the range of 293 mdpl to 1,320 mdpl and are in a hilly area with a gentle to steep slope, so the chance of erosion on open land in the rainy season, except for forest areas that are dense with annual crops, the chance of erosion is quite small.

The actual conditions obtained from the results of matching data in the field and analysis in the laboratory with land suitability tables for clove plants are S3eh or appropriate marginal with constraints on, erosion hazard (eh) for SPL 1, 2, 3 and 4 while after there are efforts to improve the potential is S2wa, eh for spl 1 while spl 2 and 4 are S2wa, eh, lp or quite in accordance with constraints on water availability (wa), erosion hazard and land preparation while for SPL 3 the results obtained are S2wa, rc, eh, lp or quite in accordance with constraints on water availability (wa), rooting media (rc), erosion hazard and land preparation. In the actual condition, it is known that there are several limiting factors described, including factors that can be repaired or limiting factors that cannot be fixed/permanent. The limiting factors that can be improved on the three land conditions are nutrient retention (nr) and erosion hazard (eh) while the limiting factors that cannot be fixed or permanent on the three critical lands are water availability (wa), root media (rc), and preparation land (lp). Cloves are an alternative choice between durian and coffee, even though they have an S3 value or marginally appropriate, but are still preferred by the community in terms of the economic value of durian, the selling price is more profitable and can improve the standard of living of farmers in the forest area after that only cloves and coffee.

Coffee plants at spl 1, 2, 3 and 4 from land suitability analysis for medium to highland areas with coffee plants as the main mainstay besides cloves and durian actually obtained S3eh results or marginally suitable with constraints on erosion hazard (eh). The research area mostly has a slope between 17.6% to 28.58% so that there is a chance of erosion hazard.The potential suitability of the land appears overall for spl 1, 2, 3, and 4 coffee plants which are the main commodities which have limiting factors that can be improved, for example, nutrient retention (nr) by applying organic matter, and erosion hazard (eh) by making terraces. Terrace. Permanent barriers cannot be repaired because they are related to land preparation or to the condition of the soil solum (Syekhfani, 2014). Coffee is in second place because it is the prima donna for the medium to highlands along with intercropping between pine and coffee to increase the income of farmers and protect production forests from looting, the benefits are when annual plants are immature while at the same time preventing land degradation due to erosion, the second Based on the survey results, coffee is the last order after cloves for all land use units.

Durian plants for all land use units from the medium to the highlands grow well and are adapted to the characteristics of the land and the characteristics of the durian plant. Low productivity or a decrease in crop yields can be caused by several factors, namely a decrease in soil fertility, an agro-climate that changes rapidly or soil management is not in accordance with soil conservation rules so that the process of soil quality decline occurs due to a lot of erosion and loss of nutrients due to the washing process. that goes into the ground.

Based on the results of the analysis showing that the limiting factor is increasing, it can be identified that land is experiencing an accelerated degradation process if land management is still carried out without land protection efforts, it is very necessary to handle land improvement and maintenance in an integrated manner to maintain fertility, capability, and productivity of land with Seeing all aspects of the problem, especially land slopes above 17.6%, the results of erosion research have begun to appear.

Evaluation of potential land suitability for 3 commodities including durian, clove, and coffee, most of the results were quite suitable (S2) with the limiting factor in land preparation (lp), especially surface rocks, which were more than 5% and rock outcrops were more than 5%, because the slope is more than 17.6% % and erosion often occurs, rock outcrops at SST 2,3 and 4 as a whole are more clearly visible and a lot of topsoil is lost on open lands for forest lands that experience very little erosion. slowly need improvements by making terraces so that the rate of erosion can be reduced.

4. Conclusion

1. Regional characteristics describe various information including morphology, physics and soil chemistry for durian, clove and coffee plants. The land suitability class is actually in accordance with marginal (S3eh) with a limiting factor on erosion hazard while for potential after undergoing improvements that are not permanently permanent in nature. Overall, the score is quite appropriate (S2) with various limiting factors but most of the constraints are on erosion hazard and land preparation. The Wonosalam area mostly has alfisol soil types, mountainous and hilly physiography, as well as volcanic ash as parent material. Alfisol soil type has a good enough ability to be cultivated for plantations and production forests.
2. The direction of forest area development in future research areas based on land suitability is an integrated farming forestry area with a technological approach, and plant suitability so that this area can become a center for research, innovation, education, training, empowerment and community assistance without compromising its function. forests as environmental conservationists in an integrated manner.

REFERENCES

Charnsungnern Maruek, Sittichai Tantanasarit, 2017, Environmental sustainability of highland agricultural land use patterns for Mae Raem and Mae Sa wat.

Datta, P.K. 2014. Agricultural Sustainability and Economic Development: Across–country analysis, *Merit Research Journal of Agricultura lScience and Soil Sciences*, 2(1):007-023.

FAO. 1983. Guidelines Land Evalution for Rainfed Agriculture. Soil Resources Management and Conservatiom Service Land and Water Development Division FAO Soil Bulletin. FAO-UNO, Romeo. No 52: 249 pp.

Gupta, Priyank., Javed Alam, Mohammad M., 2016, Influence of thickness and position of the individual layer on the permeability of the stratified soil, Perspectives in Science (2016) 8, 757—759.

Rossiter, D.G., 1994. Land Evaluation lecture notes. part 4, economic land evaluation. College of Agriculture and Life Sciences. Dept. of Soil, Crop & Atmospheric Sci., SCAS Teaching Series T94-1.

Van Noordwijk M and Hairiah K. 2006. Agricultural Intensification, Soil Biodiversity and Agro-ecosystem Function. Agrivita 28:0126–0537.

The Use of Pest Mobile Application for Optimizing the Sustainability Support of Pest Management in Oil Palm Plantation

Henny Hendarjanti[1*] and **Sapto W. Indratno**[2]

[1]PT. Astra Agro Lestari Tbk. Jl. Pulo Ayang Raya Blok OR/I Pulogadung Industrial Area, East Jakarta 13930. Tel. +62811929547
[2]Bandung Institute of Technology, Statistics Research Group, Jalan Ganesha 10, Bandung, Indonesia

*henny.hendarjanti@gmail.com

Abstract. The road map of the Indonesian palm oil industry, which is entering the industrial revolution 4.0, is by innovation and sustainable development of digitalization, referring to the application of machine automation and human-machine communication. As a significant player in the global palm oil industry, Indonesia has started to apply digital technology in the oil palm plantation industry. In this paper a digital early detection of pests and its contributions to sustainable pest control is explored. This technology will make it easier to monitor activities on the plantation on a regular basis. Real-time observation is needed to find out the problem as early as possible. This collected digital information is then statistically analyzed and linked to ecosystem services to get better decisions, especially for sustainable integrated pest control (IPM). The use of smartphone technology with an HDI implementation assessment approach and the Triple Bottom Line is the company's responsibility to the environment (planet), social (humans), and profit.
Keywords: pest management, implementation IPM, digitization, oil palm, mobile application

1. Introduction

Sustainable development, including sustainable agricultural development, is a commitment of world countries that must be obeyed and implemented. The Sustainable Development Goals (SDGs) are a global action plan agreed by world leaders, including Indonesia, to end poverty, reduce inequality and protect the environment. The SDGs contain 17 Goals and 169 targets expected to be achieved by 2030. The sustainable development approach is essentially a development activity that combines economic, social, and environmental aspects. The impact of past development, which only emphasizes economic progress, has caused social problems and environmental damage. This activity is reflected in the fact that there are still many problems related to environmental damage and the degradation of natural resources. Issues in implementing sustainable agricultural development are still common, especially in developing countries, including Indonesia. Sustainable development, including sustainable agricultural development, is a commitment of world countries that must be obeyed and implemented.

World leaders agreed on a sustainable development action program in Agenda 21 at the Earth Summit in Rio de Janeiro in 1992. This action carries the concept of multidimensional sustainability by considering achieving the objectives of the ecological, social, and economic dimensions. These three dimensions have a very close relationship and dependence. Agenda 21 contains aspects related to agriculture, such as direct (Chapter 14), Sustainable Land Management (Chapter 10), Safer Use of

Toxic Chemicals (Rivai and Anugrah, 2021). Ecologically sustainable is an activity to maintain the integrity of the ecosystem, the carrying capacity of the environment, and preserve natural resources, including biodiversity. Integrated Pest Management (IPM) is a widely used strategy to promote the wise use of pesticides. The IPM helps reduce the impact of chemical pesticides, which impact environmental damage and human health. The IPM emphasizes an early sampling system in determining pest status. Control decisions are made through the critical threshold of the pest population. The IPM is used to detect plant-disturbing organisms on time to implement appropriate controls specifically. However, operational complexity makes sampling activities difficult for practitioners following complex procedures to effectively adhere to the unpredictable and non-uniform distribution of plant-disturbing organisms, as they may occur under field conditions (Shah and Razag, 2021). In addition, the level of accuracy of the activities carried out can affect the overall use of pesticides between experienced and inexperienced practitioners.

Continuous innovation and development of information technology affect various industrial and agricultural sectors (Zambon et al., 2019). The use of digital technology in Ali's Research (2020) and Bayas et al.'s research, 2020 have made a direct contribution to the United Nations Sustainable Development Goals by advancing their implementation or providing local data systems to measure and monitor certain indicators such as the proportion of productive agricultural area and sustainable. With the same spirit, the use of computer and internet technology, such as mobile applications connected to the website, has been carried out in the trade and agricultural sectors in Indonesia.

In this paper, a short technical note detailing the technology behind the use of a mobile app in the early detection of pests in oil palm plantations is presented. This mobile app helps plantation workers easily track and monitor activities on the plantation on a regular, real-time basis to find out exactly which areas are experiencing problems. The data collected, then, being analyzed statistically and relating it to ecosystem services to get better decisions, especially for integrated pest management (IPM) in a sustainable manner. This mobile app is an approach to assessing the implementation of the IPM and Triple Bottom Line, which is the company's responsibility to the environment (planet), society (humans), and profit.

2. Material and Methods

2.1. The Mobile and Web Early Warning System

Currently, the observation of plant-disturbing organisms in oil palm plantations is using pest mobile application technology. This technology uses a gadget that has been programmed to store observation data on the condition of plant-disturbing organisms, natural enemies, and beneficial plants as habitats for beneficial insects such as parasitoids and predators. The early warning system (EWS) officer in the field will input each observation object technically following the structure and model in the Mobile Early Warning System (EWS) application. After completing all observation objects, the data in the stored device is transferred to a flash on the go (OTG). The data in this OTG flash disk is transferred to a USB port via a computer that is directly connected to the web link: ews@astra-agro.co.id in the head office via satellite. Figure 1. provides an overview of the current structure of the Mobile and Web Early Warning System applications. The EWS-Mobile produces several reports of observations that can be used by field assistants, site managers, and practitioners such as: plan and realization of EWS, daily monitoring of EWS observations, monitoring of observations of oil palm leaf-eating caterpillars, monthly reports on attacks by plant-disturbing organisms, monthly reports on the condition of biological agents and conservation of beneficial plants, monthly performance planning, daily performance planning, and others. The EWS-Web also shows some statistical analysis, such the time series model, descriptive statistics, the correlation between healthy larvae of lepidopteran pests and parasitic larvae, etc., sampling coordinates, spatial plots, and temporal & spatial dynamics pattern of lepidopteran pests (Hendarjanti et al., 2017), which can be used by the field assistants, site managers, and practitioners in carrying out the plan-do-check-action (PDCA) process.

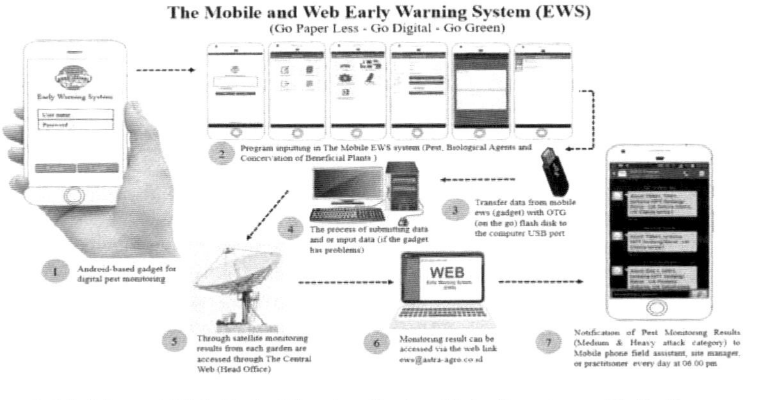

Figure 1. Mobile and Web Early Warning System Data Structure with Its Component for Applied System

3. Results and Discussion

In implementing this study, we took the case of the bagworm Clania Tertia (Lepidoptera: Psychidae) attack on the routine detection period of block E-16 PT. Ekadura Indonesia in Kunto Darussalam, Rokan Hulu Regency, Riau. The time series model (Fig.1.) shows the amount of bagworm data in the OE block 16 for five years or 60 months (2015-2020). The pattern of development of bagworms occurred in the period November 2018 - March 2020. The increase in the population of *Clania tertia* in December 2019 and the treatment of insecticide applications were shown from the information on the number of larvae exposed to insecticides. This information is a concern for management to introduce beneficial insects. The non-constant variance is indicated by an outbreak in the number of bagworms reaching about 2200, which occurred in December 2019.

The descriptive statistics table given in Fig.3. shows some information on the number of healthy bagworms *Clania tertia* in the period of March, 3rd 2020. Here the average value of all sample points with an average bagworm reaching 8.67 and a standard deviation of 7.01. In 60 months, the population of the *Clania tertia* bagworm was in the range of 0 to 26 individuals. The variance value of 49.2 and standard deviation of 7.01 shows that the number of bagworms in each month is quite heterogeneous in the range between 0-26 bagworms. The median (middle) and mode show that not all sample points contain bagworms every month. There is information on several months for each sample point with a high average value. Values for that month are categorized as upper outliers. The existence of the extreme upper value causes a very striking difference. This information is obtained by comparing the mean, median, and mode values. Each intense month has a very high number of bagworms, exceeding the habit in other months.

Correlation between healthy caterpillar populations and influencing variables is also given in the Web-EWS. The correlation shows a positive linear relationship (correlation coefficient: 0.53) which means if the population of healthy bagworms increases, the number of caterpillars exposed to insecticides also increases. It can be seen especially in November 2018 - March 2020, where the increase in healthy bagworms tends to have a longer month span than the increase in parasitic caterpillars. This correlation shows that the use of insecticides in suppressing the population of bagworms *Clania tertia* also causes a decrease in the population of beneficial insects. It was indicated by the absence of a relationship between healthy *C. tertia* larvae and pathogen-infected, parasitized, and predator-prey. This information shows that parasitoid and predatory insect populations have not been found in the area, so management needs to introduce these beneficial insects. Based on Early Warning System Mobile Data information, the coverage of valuable plants is still shallow.

Coordinate sampling plots, given in Figure 4, describe the distribution of pest populations within a block in the oil palm landscape. Based on population distribution information, we can see the proportion of pest distribution based on the criteria for the number of Bagworms populations classified as mild, moderate, and severe attacks. Threfore, we can determine whether the pest observers have done an excellent job. Pest observations should ideally follow a predetermined sample tree. If the observers do not follow the pattern sampling pattern, the results will be seen in the Sampling Image of the plot coordinates. Sampling plot coordinates will assist Management in identifying area conditions with moderate and severe pest populations and analyzing factors that affect the abundance of an organism in the oil palm landscape. Spatial plots generated from the mobile and web EWS systems, given in Figue 5, are equipped with coloring gradations ranging from green, yellow, and red. This coloring shows the level of the pest population. Figure 6 shows the dynamic behavior of the bagworm population and the emergent properties of the interacting systems. Temporal and spatial dynamics of the bagworm population from July 2019 to June 2020 show the dynamics of the bagworm population in Safe conditions in August and December 2019, the dominant attack condition in the Mild category in July, September, October, November 2019, and April - June 2020. February - March 2020 has shown an increase in the population dynamics of bagworms in the heavy category. In January 2020, the bagworm population began to show an increase in the Bagworms population from the Light class to some Moderate and Heavy. Based on the instructions of The Temporal and Spatial Dynamic Pattern of Bagworms *Clania tertia*, it is hoped that the management can immediately take the proper control strategy decisions before the Bagworms population moves towards a critical population as happened in February - March 2020. Management must be keen to analyze and see what opportunities are causing it. Two months (August and December 2019) safe condition event. Through The Temporal and Spatial Dynamic Pattern of Bagworms *Clania tertia*, it is hoped that the management can set Integrated Pest Management (IPM) strategy by including biological control components as a priority. The use of chemicals in control should be the last alternative.

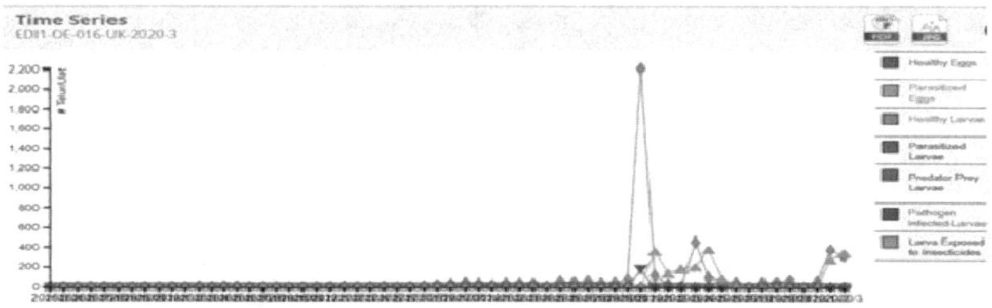

Fig. 2. The Time Series of Bagworms Population

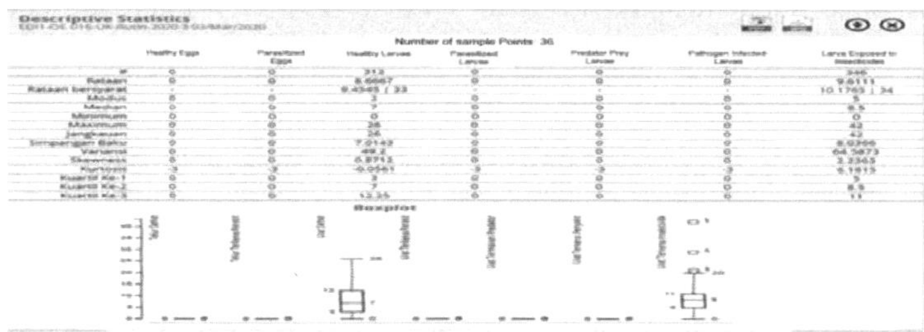

Fig.3. The Descriptive Statistics of Bagworms Population

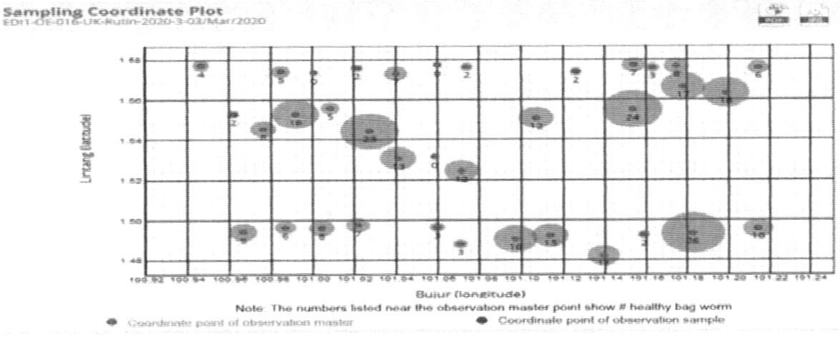

Fig. 4. The Sampling Coordinate Plots of Bagworms Population

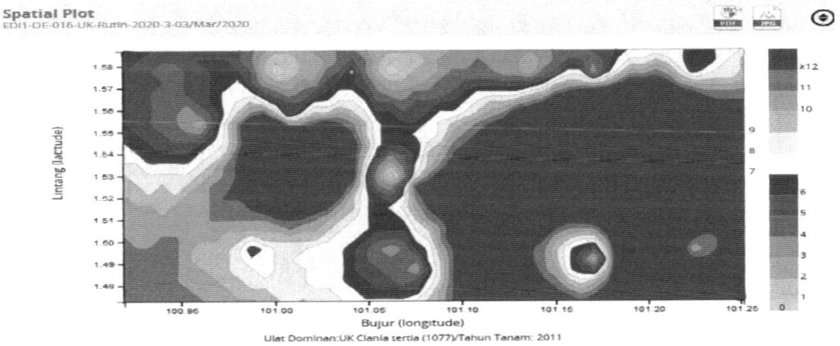

Fig. 5. The Sampling Coordinate Plots of Bagworms Population

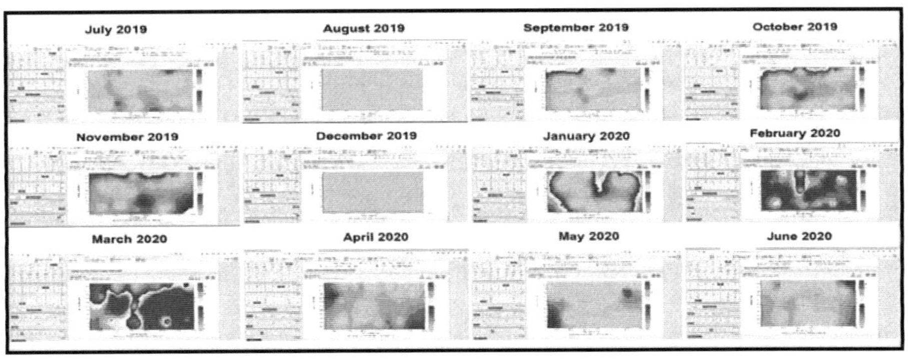

Fig. 6. The Temporal and Spatial Dynamics Pattern of Bagworms

4. Conclusion

Monitoring pests early and carried out continuously is s needed to determine the abundance of pests and beneficial insects as natural enemies in a landscape. Monitoring is necessary to resolve manyproblems and beneficial insects as natural enemies. Mobile and Web Early Warning System (EWS) is an automation system in carrying out pest detection that can change the information process from analog to digital form. The benefit for management is the speed of information on the distribution and condition of pest attacks. The digitalization system through mobile and web EWS provides information about the role of natural enemies and the carrying capacity of beneficial plants as a habitat for beneficial insects. The presence of pest populations in the landscape becomes for evaluation and analysis in determining pest control policies. The management focuses on environmentally friendly controls to sustainability corporate. The prioritizes biocontrol control pest with the involvement of biological agents as an ecosystem service. Using the IPM principle, the control must be carried out when the pest population density is at an economic threshold not to exceed the Economic Injury Level. Chemical control is the last alternative after biological, physical, and mechanical control.

5. References

[1] Rivai, R.S., and Anugrah, I.S. 2011 Concept and Implementation of Sustainable Agricultural Development in Indonesia. Forum Penelitian Agro Ekonomi. 29(1): 13 – 25.

[2] Shah F.M.,and Razaq M. (2021) From Agriculture to Sustainable Agriculture: Prospects for Improving Pest Management in Industrial Revolution 4.0. In: Hussain C.M., Di Sia P. (eds) Handbook of Smart Materials, Technologies, and Devices. Springer, Cham. DOI: 10.1007/978-3-030-58675-1_76-1.

[3] Zambon, I., Cecchini, M., Egidi, G., Saporito, M.G., and Colantoni, A. 2019. Revolution 4.0: Industry vs. Agriculture in a Future Development for SMEs. *Processes*. 7(1):36. DOI: 10.3390/pr7010036.

[4] Ali, M. 2020. Intention Level of Farmers to Use Information Communication Technologies for agricultural Risk Management in Malaysia. Journal of International Agricultural and Extension Education. 27(2): 108-117. DOI: 10.5191/jiaee.2020.272108.

[5] Bayas, J.C.L., Gardeazabal, A., Karner, M.,Folberth, C., Vargas, L., Skalský, R., … Govaerts, B. 2020. Agro Tutor: A Mobile Phone Application Supporting Sustainable Agricultural Intensification. Sustainability 12: 2-10. DOI:10.3390/su12229309.

[6] Panda, C. K. and Egbuna, C. (Eds.). 2020. Advances in Application of ICT in Crop Pest and Disease Management. Natural Remedies for Pest, Disease and Weed Control,235–242. DOI: 10.1016/B978-0-12-819304-4.00020-8.

[7] Nurhayati-Wolff, H. 2021. Smartphone users in Indonesia 2017-2026. Retrieved October 31, 2021, from https://www.statista.com/statistics/266729/smartphone-users-in-indonesia.

[8] Hendarjanti, H., Indratno, S.W., Ismunandar., U.T. and Sugeng W., M.H. 2018. The potency of an automated detection system for real-time pest monitoring from smartphones and spatial analysis for leaf-eating caterpillar attack evaluation. Earth and Environmental Science 183: 1-7. DOI: 10.1088/1755-1315/183/1/012021.

[9] Wood, B.J. and Kamarudin, N. 2019. Bagworm (Lepidoptera: Psychidae) Infestation in The Centennial of The Malaysian Oil Palm Industry- A Review of Causes and Control. Journal of Oil Palm Research. 31(3): 364 -380. DOI: 10.21894/jopr.2019.0032.

[10] Damos, P. 2016. A stepwise algorithm to detect significant time lags in ecological time series in terms of autocorrelation functions and ARMA model optimization of pest population seasonal outbreaks. Stoch Environ Res Risk Assess 30:1961–1980. DOI: 10.1007/s00477-015-1150-1.

[11] Tsai, C.T., Sung, F.C., Chen, P.S., and Lin S.C. 2011. Exploring the spatial and temporal relationships between mosquito population dynamics and dengue outbreaks based on climatic factors. Stoch Env Res Risk Assess. 26:671–680.

[12] Lin C.H., Wen, T.H., Teng, H.J., & Chang, N.T. 2014. The spatio-temporal-characteristics of potential dengue risk assessed by *Aedes aegypti* and *Aedes albopictus* in high epidemic areas. Stochastic Environmental Research and Risk Assessment. 30(8) DOI:10.1007/s00477-014-0940-1.

AUTHOR INDEX

Institute of Physics
Dirac House, Temple Back
Bristol BS1 6BE UK

ISSN: 1755-1307
ISBN 978-1-7138-6988-7